Mosquito
Diversity and Control

The Authors

Dr. Sathe Tukaram Vithalrao [M.Sc., Ph.D., Sangit Vishard, IBT (Seri.), F.I.S.E.C., F.S.E.Sc., F.S.L.Sc., F.I.C.C.B., F.S.S.I.] is presently working as Professor and Head, Department of Zoology, Shivaji University, Kolhapur. He has teaching experience of 28 years in Entomology at University PG department and 15 years in Agrochemicals and Pest Management. He has written 30 books and published 255 research papers in national and international journals of repute. He guided 18 Ph.D. students and completed 4 major research projects (from CSIR, DST and UGC). He visited Canada (1988), Japan (1988), Thailand (2002, 2004), Spain (2005), France (2005), South Korea (2006) and Nepal (2007) etc. for academic work. He is member of editorial board of eleven prestigious journals. He delivered 30 talks through All India Radio and internal conferences and involved in Doordarsha, S.T.V. and B. T.V. programmes on biodiversity of moths and butterflies. He published more than 35 popular articles in daily newspapers on insects and sericulture. He got several prestigious awards like "Environmentalists of the Year-2003", "Bharat Jyoti", "Jewel of India", "International Gold Star", "Eminent Citizen of India", "Education Acumen", "Best Educationist", "Eminent Scientist of the Year-2008", "Lifetime Education Achievement" and "Lifetime Achievement in Entomology and Insect Taxonomy-2009", Educational Leadership, Asia Pacific International Award etc. He is also working as Research and Recognition (RR) Committee member for Pune University, Pune; North Maharashtra University, Jalgaon; Shivaji University, Kolhapur and R.R. and Faculty of DBA Marathwada University, Aurangabad. He has been also awarded several fellowships from different scientific and academic societies. He is Chairman of Maharashtra District Environmental Centre of NESA.

Dr. Mahendra B. Jagtap (M.Sc., Ph.D.) is zonal entomologist, National Vector Borne Disease Control Programme, Satara (M.S.) since last 15 years and Assistant Director Health Services (M), Kolhapur. He has published 20 research papers in the journals of International repute. He is convenor of NESA district environmental centre, Satara. He has contributed as co-author for Mosquito Borne Diseases alongwith Dr. T.V. Sathe and Dr. Aswari Sathe.

Mosquito Diversity and Control

T.V. Sathe
Head
Department of Zoology
Shivaji University
Kolhapur – 416 004, M.S.

Mahendra Jagtap
District Maleria Centre
Satara, M.S.

2013
Daya Publishing House®
A Division of
Astral International Pvt. Ltd.
New Delhi – 110 002

Published by   :   **Daya Publishing House®**
                          A Division of
                          Astral International Pvt. Ltd.
                          -ISO 9001:2008 Certified Company
                          4760-61/23, Ansari Road, Darya Ganj
                          New Delhi-110 002
                          Ph. 011-43549197, 23278134
                          E-mail: info@astralint.com
                          Website: www.astralint.com

Laser Typesetting   :   **Classic Computer Services**
                          Delhi - 110 035

Printed at   :   **Salasar Imaging Systems**
                          Delhi - 110 035

PRINTED IN INDIA

Preface

Mosquitoes are important vectors for many diseases such as malaria, filariasis, Japanese encephalitis, yellow fever, dengue, chikungunya etc. They belong to Order – Diptera, family Culicidae and are estimated about 3500 in species number all over the world. So there is lot of confusion in identification of species and disease spreading. Therefore, any advance knowledge on taxonomy, seasonal abundance and distribution of mosquitoes have practical utility in adopting their appropriate control measures solving the problems related to health care.

Molecular techniques have proven very useful for identification of particular genotypes and will lead to major medical advances within the coming years against mosquito borne diseases, tuberculosis and HIV/AIDS. Hence, in the present text, attempts have been made on taxonomical descriptions and seasonal abundance of mosquitoes in Western Maharashtra (Districts : Kolhapur, Sangli, Satara and Pune). In all, 31 species of mosquitoes have been reported from the region of which 21 species were common and 10 species were rare. In the text, 13 species of mosquitoes were newly described and 3 redescribed. Hence, we feel that the present work will be stimulatory and useful for students, teachers, officers and scientists in entomology and epidemiology.

T.V. Sathe

Mahendra Jagtap

Contents

Genus: *Anopheles* **Meigen**

Anopheles (Anopheles) *compestris*

Anopheles (Anopheles) *kolhapuri* sp. nov.

Genus: *Aedes* **Meigen**

Aedes (Stegomyia) *aegypti*

Aedes (Stegomyia) *albopictus*

Aedes (Mucidus) *sathei* sp. nov.

Aedes (Finlaya) *rajashri* sp. nov.

Genus *Culex* **Linnaeus**

Culex (Culex) *quinquifasciatus*

Culex (Culex) *malhari* sp. nov.

Culex (Culex) *malkapuri* sp. nov.

Culex (Barraudius) *mirjensis* sp. nov.

Culex (Barraudius) *satarensis* sp. nov.

Anopheles (Cellia) *subpictus*

Anopheles (Cellia) *fluviatilis*

Culex (Culex) *vishnui*

Chapter 1
General Introduction

Insects with their great diversity in form, colour and habitat specialization together make them one of the most rewarding groups of animals to study. They exhibit seasonal and geographical variations and interact with living environment in several ways. However, adverse conditions generated by man in the environment such as deforestation, monoculture practices, excessive use of pesticides, habitat modifications etc. often cause biodiversity crisis and the insect ecology and population dynamics often reflects these anthropogenic interferences.

The type of insect pest capable of disseminating pathogenic diseases from one mammalian host to another or one plant to another is called vectors. Several genera of arthropods play a role in human disease, but mosquitoes are the most notable disease vectors. The most significant mode of vector borne disease transmission is by biological transmission through blood feeding arthropods. The pathogen multiplies within the arthropod vector and the pathogen is transmitted into host when the arthropod takes a blood meal. Mechanical transmission of disease agents may also occur when arthropods physically carry pathogens from one place or host to another, usually on body parts. Majorities of vector borne diseases survive in nature by utilizing animals as their vertebrate hosts and are therefore, zoonoses. For a small number of zoonoses, such as malaria and dengue, human are the major host, with no significant animal reservoirs. Following are the key components that determine the occurrence of vector borne diseases.

(1) The abundance of vectors and intermediate and reservoir hosts.

(2) The prevalence of disease causing pathogens suitably adapted to the vectors and the human or animal hosts.

(3) The local environmental conditions especially temperature and humidity.

(4) The resilience behavior and immune status of human population.

Dipterans dominate as vectors of a number of human and livestock diseases over other members of the orders. Mosquitoes profoundly play an important role making enormous impact by spreading epidemics. The disease outbreaks are due to both biological and environmental factors that encourage vector breeding. Therefore, to study the relationship between humans, vectors and pathogens with respect to disease transmission is very crucial.

The mosquitoes belong to the phylum Arthropoda, Order Diptera, Sub order Nematocera and the family

Culicidae. According to Knight and Stone (1977) there are 3,220 species of mosquitoes in the world belonging to 34 genera, out of these India represents 16 genera (Barraud 1934).

Mosquitoes are small, slender, two winged insects covered with hairs and scales, and with three body divisions *viz.* head, thorax and abdomen. Prognathus head contains a pair of compound eyes and densely hairy with 14-15 segmented antennae, plumose in males and pilose in the females. Proboscis is long, brown or black modified for piercing and sucking the blood. Palpus is about the length of the proboscis in the males but, in females it differs in length in the different genera. Thorax is convex, made up of pro, meso and metathorax and externally undifferentiated. It consists of pair of wings and pair of halters. Thorax carries three pairs of legs. Each leg consists of coxa, followed by femur, tibia and tarsus. Tarsus is composed of five joints. Each leg ends with a pair of claw. Abdomen, the last division of body consists of eight segments and terminates in a pair of claspers in males and in the female as lobed appendages.

Primarily males and females of mosquito species feed on nectar and other plant juices. However, the females of some species need vertebrate blood to produce eggs. The females thus have very specialized long, piercing and sucking mouth parts. By piercing the skin of their prey they obtain the blood feed. Individual female mosquitoes can lay eggs in several batches in their life span. However, a blood meal is essential for each batch. The females of different species feed at different times of the day and in different types of locations within the same surroundings. Any vertebrate animal may provide the required proteins to mosquitoes. The mosquitoes have specific prey feeding preferences.

The life cycle of mosquitoes passes through four distinct stages *viz.* egg, larva, pupa and adult (Busvine, 1980). Eggs are laid either singly or in clusters. The cluster may contain about 300 eggs attached together to form a soft of 'raft'. Some mosquito species lay eggs on the surface of calm water, while others lay on damp soil destined to be flooded either way, the eggs end up floating at the water's surface. Incubation period is 2-3 days. In some cases eggs laid in late season may withstand the harsh conditions of winter and hatch in the following spring (Busvine, 1980).

Larval period of mosquito varies from 4-14 days in the water. They breathe air through a tube to the surface and eat microorganisms and other organic matter from water. The larvae metamorphosed into pupae (tumblers). The pupal stage is non feeding stage. Pupae float at the surface but are capable of diving/tumbling to safety when disturbed. The pupal period is about 1-4 days. Fully developed adult mosquito breaks the pupal case and comes out to the surface of the water, where it rests until its body dries and hardens. After two days of emergence mating and feeding occurred.

Temperature affects species specific characteristics and the rate of development. Under favorable warm conditions, some mosquitoes develop from egg to adults in less than a week. Adult female mosquitoes mostly survived for about two weeks. They may parasitize humans or predated by birds, bats, amphibians or spiders. However, adult females may hibernate throughout the winter if emerge late in the season. They start egg laying in the spring.

Mosquitoes are responsible for spreading various fatal diseases like malaria, filariasis, dengue, chikungunya, encephalitis, yellow fever etc. Fowl pox of poultry, mycomatosis of rabbits, rift valley fever of sheep, encephalitis of horses and birds, heat worm of dogs etc. are also transmitted from animal to man and some diseases from other insects. South American warble fly *Dermatobia* sp. transports its eggs to the skin of man and animals through mosquitoes alone and cause myiasis on skin after hatching (Sathe and Tingare, 2010).

The human malarial protozoan pathogen refers to *Plasmodium vivax, Plasmodium falciparum* (malignant tertian, subtertian aestivoautumnal malaria, falciparum malaria), *Plasmodium malaria* (quarten malaria) and *Plasmodium ovale* (ovale malaria). Races and strains are visualized in case of *P. vivax* and *P. falciparum.* They are confirmed through the clinical picture, geographical distribution and immunological responses. *P. vivax* causes 65-90 per cent malaria while, *P. falciparum* 25-30 per cent of the total malarial cases (Sathe and Girhe, 2001). *P. malaria* shows only 1 per cent cases but, *P. ovale* is reported sporadically. The protozoa invade the parenchyma cells of the liver and after passing through development stage, attach and resides inside R.B.Cs. Female mosquito inject saliva and sporozoites during the act of biting and feeding. Other means of malarial transmission can be either through blood transfusion or congenital transmission as the use of a syringe from an infected person. (Sathe and Tingare, 2010).

The species of the genus *Anopheles* specially *Anopheles labranchiae* Flalleroni, *An. sacharovi* Favre, *An. sargentii* (Theobald), *An. superpictus* Grassi and *An. pharoensis*

Theobald are reported from parts of the Mediterranean area, *An. funestus* Giles, *An. moucheti* Evans, *An. nili* (Theobald) and member of the *An. gambiae* Giles complex in the Ethiopian region, *An. stephensi* Liston, *An. flaviatilis* James and *An. pulcherrimus* Theobald from Western Asia *An. culcifacies* Giles from India and Ceylon, while, *An. maculatus* Theobald, *An. sundaicus* (Rodenwaldf) and member of the *An. barbirostris* Vander Wulp, *An. hyrcunus* (Pallas) Wulp group are reported from Southern Asia. The members from *An. punctulatus* complex in Melanesoa and *An. pseudopunctipennis* Theobald, *An. bellator* Dyar and Knab, *An. cruzi* Dyar and Knab, *An. darlingi* Roof, *An. aquasalis* Curry, *An. albimanus* Wiedemann, *An. albitarsis* Lynch Arribalzga and *An. nuneztovavi* Gobaldon are reported from Central and South America (Smith, 1969).

The Indian species of *Anopheles* transmitting malaria refer to *An. annularis* Van der Wulp, *An. sundaicus* RodenWaldt, *An. stephensi* Liston, *An. maculatus* Theobald, *An. philippinensis* Ludlow, *An. leucosphyrus* Doenitz, *An. fluviatilis* James, *An. culicifacies* Giles, *An. varuna* Iyengar and *An. minimus* Theobald (Rao, 1984). The dominant species of *Culex* includes - *Culex pipiens* Linnaeus, *C. vishnui* Theobald, *C. pseudovishnui* Colless, *C. tritaeniorhynchus* Giles, *C. bitaeniorhynchius* Giles, *C. sinensis* Theobald, *C. gelidus* Theobald, *C. sitiens* Wiedemann while, the *Aedes is* represented by *Aedes aegypti* Linnaeus, *A. albopictus* Skuse, *A. vittatus* Bigot and *A. variegatus* Schrank mostly.

Filaria pathogens in man are two nematodes viz, *Wuchereria bancrofti* and *Brugia malayi*. The adult worms live in the lymphatics, produce live embryos (microfilariae) which invade the blood stream. The embryos do not develop further in the blood. According to Sathe and

Tingare (2010) a mosquito vector imbibes microfilariae along with the blood and develops further in 10-12 days, depending upon the environment. The third stage infective larvae are deposited on the human skin by the mosquito when it visits man for the next food. Such larvae enter through the wound. The minimum time taken from the entry of the third stage infective larvae until the microfilaria first appear in the peripheral blood is about 82 days, but it may take as long as one year. The female nematodes begin to liberate microfilaria which invades the blood stream. The cycle is thus repeated.

It is estimated that at least 250 million people are infected with *W. bancrofti* and *B. malayi* from the world. About one billion people from tropical and sub-tropical countries are at risk and about 200 million are actually suffering from the disease. In India more than 200 million populations is at risk. From the world about 90 species of mosquitoes have been recorded as natural vectors of *W. bancrofti*. *Culex pipiens* Linnaeus is the vector in India. *Anopheles gambiae* A and B and *Anopheles funestus* are important vectors of filaria in rural East Africa.

Coastal parts and banks of rivers are distributional places of *Bancroftian filariasis* in South India. The most favorable period of transmission is during the monsoon but found throughout the year. The favorable conditions for vector and parasite are 15.5°C-32.2°C and at least 60 per cent relative humidity (RH).

Brugia malayi cause *Malayian filariasis* from Orissa, Madhya Pradesh, Tamil Nadu, Kerala, West Bengal and Assam. The mosquitoes of the *Mansonia* group and *Anopheles barbirostris* are the vectors in India. Important reservoirs for this worm are monkeys (*Peresbytis obscurus*), cat, dog and some other animals.

JE (*Japanese encephalitis*) is transmitted by *Culex* mosquitoes. The pathogen involved is Virus (JEV) which is maintained in nature by a complex cycle that involves pig as amplifying host arderid birds as reservoirs and mosquitoes as vectors. *Culex vishnui* Theobald, *Culex tritaeniorhynchus* Giles and *Culex pseudovishnui* Colless have been implicated as major vectors of JE in India. 16 species of mosquitoes are susceptible for JEV. These mosquitoes transmit viral encephalitis in man. JE is an arboviral (B) infection. The virus was first isolated in 1935 by Japanese workers.

Yellow fever is transmitted by *Aedes* and other mosquitoes an acute specific viral fever of short duration. YF virus is found in certain wild monkeys and other reservoir hosts in Africa and South America. *A. aegypti* mosquitoes bite infected men and lead mosquito-man cycle. In South America, day biting mosquitoes *Haemagogus spegazzinii, Aedes leucocelaneus* and *Sabetbes chloropterus,* are vectors and monkey-mosquito monkey cycle goes on. In West Africa, night biting mosquitoes *Aedes, Culex, Ochlerotatus, Sabethinus* and *Anopheles* transmit yellow fever virus to human beings in the forests. Infected monkeys may enter human habitations and became source of further transmission. Incubation period in man is 3-6 days (Atwal 1933).

Dengue Fever (DF) is an acute viral infection for which *Aedes aegypti* is vector. It breeds in clean water containers and elsewhere.

Rush, in 1780, called it as 'break-bone' fever. Graham in 1905 was the first to report the mosquito vector for dengue. *A. aegypti* as a vector of dengue was proved in 1906 by Bancrofti in Australia. The arbovirus (B group) is existing in four forms *viz.*, Dengue I, Dengue 2, Dengue 3 and Dengue 4, all are transmitted mainly by *A. aegypti*. Other species of the genus *Aedes* (*A. albopictus, A. squtellaris, A. albimanus* and *A. hebridus*) and *Armigeres obtarbans* are also known as vector for Dengue. *A. aegypti* maintains the disease cycle in man, while *A. albopictus* and others living in the bush or forests help in maintenance of infection among monkeys (jungle dengue). The Dengue is prevalent in tropical and sub-tropical areas of the world.

The most troublesome disease, chikungunya was first reported from Kolkatta in 1963. Its name is derived from the Swahili word meaning "that which bends up" in reference to stooped posture developed as a result of the arthralgia (severe joint pains). The first outbreak of CHK virus observed in Kolkata followed by Chennai, Pondicherry, Vellore and Vishakapattanam in 1964. Later, it was recorded from Central part of India. *i.e.* Rajmundri, Kakinada (A.P.) and Nagpur in 1965. Chikungunya incidence was high as 40 to 70 per cent in certain wards of Nagpur city in 1965. Now all the age groups are susceptible to CHK and are widely distributed throughout India.

The virus is maintained in nature at a low level in man-mosquito-man cycle. The survival of CHK virus in nature is also through transovarial transmission (TOT) in *Aedes aegypti* mosquitoes. *Aedes aegypti* is the principal vector of this virus in India. However, chikungunya can be transmitted by *Aedes albopictus, Aedes vittatus* and some *Culex* species.

When a mosquito bites to west Nile virus infected bird, it picks up viral particles circulating in the blood. Once inside the body of a mosquito they can serve as a vector, the virus co-opts the mosquito's cells into replicating the virus. After 5-14 days virus multiplies and goes to salivary glands of mosquito from where they are transported into the subsequent host when the mosquito bites for meal. West Nile Virus is primarily an avian pathogen and is transmitted among birds by ornithophilic (bird biting) mosquitoes.

Study area

Western Maharashtra (Figure 1) is leading in agriculture and industrialization and has several water bodies and hilly areas like Western Ghats. Due to Western Ghats, the geography and the climate of Western Maharashtra is more or less similar in the districts Kolhapur, Satara, Sangli and Pune (Figures 2 and 3). The districts selected for the study are shown in Figures 2 and 3.

Kolhapur district (Figures 2 and 3) is located between 15° to 17° North latitude and 73° to 74° East longitude. The district is bounded by Sangli district at the North, Belgaum district of Karnataka State at the South and East and Ratnagiri and Sindhudurg districts at East and West respectively. Kolhapur district has 7,633 sq km area with population more than 20,03,953. 12 tahsils are included under this district.

Sahyadri mountains in the west, provides several spur's in the East of the district. Major portion of the district is 390 to 600 meters above mean sea level. Krishna, Warna, Panchaganga, Doodhganga, Vedganga and Hiranykeshi

Figure 1: Map of India Showing Maharashtra

are the principal rivers of Kolhapur district. Panchganga is formed by the five tributaries namely, Kasari, Kumbhi, Dhamana, Tulashi and Bhogawati. Panchganga merge into the Krishna at Narsobawadi in Shirol tahsil. The South Western region of the district is drained by Doodhganga river. Forests in Kolhapur district are confined to the

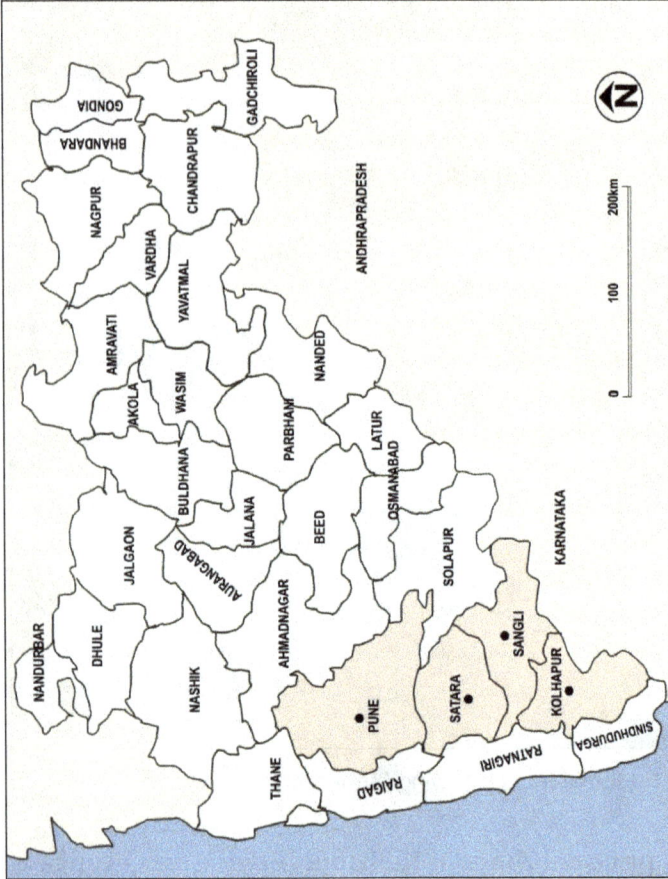

Figure 2: Map of Maharashtra Showing Study Area (Western Maharashtra: Districts Kolhapur, Sangli, Satara and Pune)

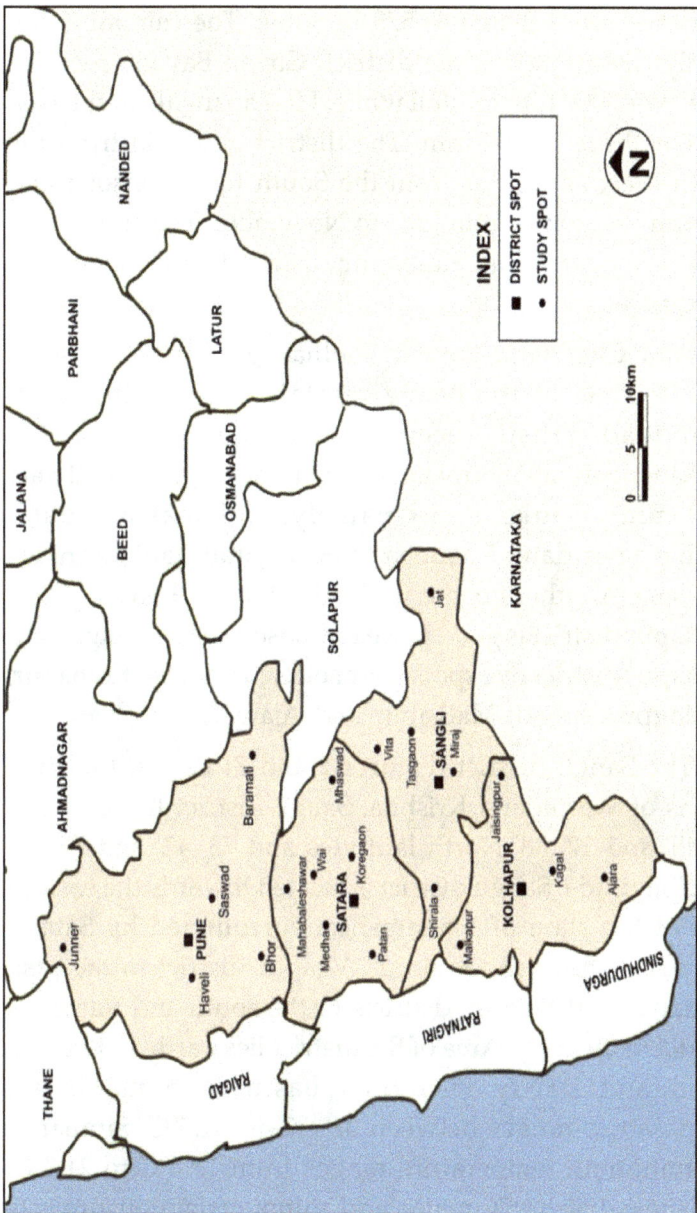

Figure 3: Map of Western Maharashtra: Study Area (Districts Kolhapur, Sangli, Satara and Pune)

Western half of the district. The total forest area in Kolhapur district is more than 1,46,575 hectares. The rainfall is not evenly distributed in the district. Gagan Bavada receives little over 6000 mm rainfall while, Hatkanangale in the East receives little as 500 mm. The district gets rain from the South West as well as from the South East monsoon and the rainy season is from June to November. The above facts clearly indicates that Kolhapur district is good place for mosquitoes.

Ajra, Chandgad, Bavada, Radhanagari and Shahuwadi tahsils come in the heavy rainfall tract. In this tract Chandgad tahsil receives 6232 mm rain while, Hatkanangale and Shirol tahsils are under poor rainfall tract (600 mm). Tanks/dams namely, Kalamawadi dam, Radhanagari dam, Kalamba tank, Rajaram tank, Rankala and several others come under Kolhapur district. Thus, Kolhapur district is good place for mosquito breeding. From Kolhapur district five spots were selected namely, Kolhapur, Jaysingpur, Kagal, Malkapur and Ajara.

The Sangli district (Figures 2 and 3) lies on the river basins of Warna and Krishna. Sangli district lies between 16°.45' and 17°.38' North latitudes and 73°.42' and 75°.40' East longitude. Sangli district is located towards the eastern part of the state of Maharashtra surrounded by Satara, Solapur districts to the north, Vijapur district to the east, Kolhapur and Belgum districts to the south and Ratnagiri district to the west. Area of the district lies partly in Krishna basin and partly in Bhima basin. The maximum temperature ranges between 31.1°C to 41.5°C. Similarly, the minimum temperature ranges from 10.3°C to 21.5°C. In Sangli district 76 major and minor irrigation projects have been launched. Therefore, mosquitogenic conditions are favorable for transmitting the vector borne diseases. From Sangli district five spots were selected namely, Miraj, Jath, Tasgaon, Vita and Shirala.

Pune (Figures 2 and 3) is well known as the 'Queen of Deccan' due to its scenic beauty, rich natural resources and good education centre. Pune district of Maharashtra has an area of 15,637 sq.km. It is lies between 18°.21' North and 73°.51' east. Pune district is located in the western part of the state of Maharashtra, bounded by district Thane to the northwest, Raigad to the west, Satara district to the south, Solapur district to the southeast Ahmednagar district to the north and northeast. The district is situated in the Western Ghats or Sahyadri mountain range and to the Deccan Plateau on the west. Average rainfall in the Western part is 3,800 mm and in the Eastern part 500 mm. It is the second largest district in the state and covers 5.10 per cent of the total geographical area of the state. April and May are the hottest months in the district. Maximum temperature during these months often rises above 36 °C. In Pune districts five spots were selected for study *i.e.* Bhor, Baramati, Junner, Saswad and Pune.

Satara district (Figures 2 and 3) is located in the western part of Maharashtra. Satara district of Maharashtra has an area of 10,492 sq.km. Which form a part of Deccan plateau between latitude 17°5' and 18°1' N and longitude 73°33' and 74°74' E. Satara district is bounded by Pune district to the north, Solapur district to the east, Sangli district to the south and Ratnagiri district to the west. Raigad district lies to its northwest. The famous Koyana dam is situated in Western Ghats of Satara districts. The Koyana basin covers West coast semievergreen forest with annual rain fall about 966 mm. Mahabaleshwar includes tropical evergreen forests. It is situated in the river basins of the Bhima and Krishna River. In Satara district seven spots were selected for study namely, Medha, Wai, Mahabaleshwar, Satara, Patan, Mhaswad and Koregaon.

On the basis of geographical and climatical parameters Kolhapur, Sangli, Satara and Pune districts were selected under western Maharashtra as a study area. Secondly, these districts have great importance in agriculture and industrialization and thirdly they contain several breeding places for mosquitoes.

Chapter 2
Review of Literature

Taxonomical

Review of literature indicates that taxonomy of mosquitoes has been studied by several workers in the world. Theobald (1901a) published a monograph on Culicidae of the world. He (1901b) erected one subgenus *Mucidus* and described one new species from Myanmar. From West Indies Coquillett (1901) described three new species of subgenus *Anopheles*. Theobald (1902) gave classification of the genus *Anopheles* while, Coquillett (1902) described six American new species of subgenus *Culex*.

In 1904a Theobald erected one more new genus *Hodgesia* of Culicidae from Uganda. Later, he (1905a) erected a new subgenus *Lophoceratomyia* and described one new species *Culex (Lophoceratomyia) fraudatrix* from New Guinea. From Ceylon James and Liston (1905) described five new species of subgenus *Culex*. Grabham (1906) described four new species of subgenus *Culex* from West Indies. From West Indies only Coquillett (1906) described five new species of subgenus *Aedes*. Banks (1906a) described four new species of subgenus *Culex* from Philippines and published a checklist of Philippines. Banks (1906b) again erected one more new genus and described ten new species of the genus *Culex* from Philippines. He also described four new species of subgenus *Banksinella* from the Philippines (Banks, 1909). Carter (1910) described one new species of *Anopheles* from South Africa. Idem (1910) published a world monograph of the Culicidae. From Uganda Newstead and Carter (1910) erected one new genus *Dactylomyia*, he described three new species of genus *Anopheles*. From Sudan Theobald (1911) erected new subgenus *Diceromyia* under which he described two new species. Carter (1911) described one new species of genus *Mansonia* from Uganda. Later, Brunetti (1912) published an annotated catalogue of Oriental Culicidae while, Edwards (1912) published key to the Australasian species of *Ochlerotatus* genus. Edwards (1913) also published new synonymy for Oriental Culicidae.

Brunetti (1914) contributed on critical review of genera in Culicidae. Banks (1914) described one new species of *Anopheles* from Philippines. Later, Strickland (1916) described one more a new species *Myzorhynchus hunteri* from Malaya. Similarly, Edwards (1916) described eight new species of subgenus *Culex* from British Museum collection.

Edwards (1920) published catalogue of Oriental and South Asiatic Nematocera and in 1921b he published a synonymic list of the mosquitoes and described key to the genera and species for Sweden. Later, Edwards (1932) reported 30 genera, 89 subgenera, 1,400 species, 146 varieties and 1,286 species synonyms. While, Evans (1931) contributed on one new subspecies *Anopheles funestus* Giles from Southern Rhodesia. Marsh (1933) added one new species of subgenus *Myzomyia* from Southwest Persia.

From Southern Europe Hackett and Lewis (1935) reported new variety of *Anopheles maculipennis*. While, Baker (1936) have been described one new species of subgenus *Orthopodomia* from China. Mulligan and Puri (1936) described *Anopheles* (Anopheles) *habibi* as a new species from Quetta, Baluchistan. Salem (1938) described two new species of genus *Topomyia* from Egypt. Similarly, Keshishian (1938) described *Anopheles sogdianus* as a new species from Tajikistan.

In 1940 Bates gave nomenclature and taxonomic status of the mosquitoes of the *Anopheles malulipennsis* complex. Belkin and Schlosser (1944) contributed on one new species of *Anopheles* from Solaman Island. Belkin (1945) added *Anopheles nataliae* as a new species from New Georgia. Stone (1945) published a mosquito's synonym. Bohart (1946) published key to the Chinese Culicidae mosquito. King and Hoogstraal (1946) described two new species of subgenus *Finlaya* from New Guinea.

De Meillon (1947) published key to the *Anophelini* of the Ethiopian geographical region. Stone and Penn (1948) erected one new subgenus and described two new species of the genus *Culex* from China. Bohart (1950) described one new species of subgenus *Orthopodomyia* from California. He also described one new species of *Culex* from Okinawa in 1953. Reid (1953) reported *Anopheles hyrcanus* group in South-east Asia.

Pringle (1954) published key to the adult *Anopheline* mosquitoes from Iraq and neighbouring territories.

Thurman (1954) visualized *Ayurakitia* new genus from Northern Thailand. Carpenter and Lacasse (1955) errected one new species of *Culex* from North America. Bohart (1956) described one new species of subgenus *Coquillettidia* from Southern Ryukyu Island. Barr (1957) described one new species of *Culiseta* from North America.

Stone *et al.* (1959, 1961, 1963, 1967, and 1970) published a synoptic catalogue of the mosquitoes of the world. They reported 31 genera, 95 subgenera, 2,401 species and 208 varieties and 1,286 species synonyms. From New Guinea Colless (1960) described four new species of subgenus *Culex*. Shahgudian (1960) gave a key to Anopheline mosquitoes of Iran. Services (1962) provided a key to the West African *Anopheline*. Baisas and Dowell (1965) published key to the adult and larval anopheline of the Philippines. Colless (1965) erected a new subgenus *Lophoceraomyia* from Malaya.

Scanlon *et al.* (1968) published an annotated check-list of the *Anopheles* mosquitoes from Thailand. Reid (1968) reported key to the *Anopheline* mosquitoes for Malaya and Borneo. Stone and Barreto (1969) erected one new subgenus and described one new species of genus *Culex* from Colombia. Later, Cagampang-Romas and Darsie (1970) also provided a published key to the *Anopheles* mosquitoes of Philippines Island. Postiglione *et al.* (1970) reported *Anopheles maculipennis* complex from Turkey. Lee and Lein (1970) gave pictorial key to the mosquitoes of Korea. White (1975) provided a catalogue of Culicidae of the Ethiopian region. Knight and Stone (1977) also provided a catalogue of mosquitoes of the world. White (1985) described *Anopheles bwambae* new species from Semliki valley, Uganda. Shidrawi and Gillies (1987) added *Anopheles paltrinierii* as a new species to Oman. While, Peyton and Ramalingam (1988) described *Anopheles* (Cellia) *nemophilous* new species from peninsular Malaysia.

Wilkerson and Strickman (1990) gave key to the female *Anopheles* mosquitoes to Central America and Mexico. Similarly, Click (1992) published key to the female *Anopheles* mosquitoes from Southwestern Asia and Egypt. Ward (1992) published a catalogue of the mosquitoes of the world. Zaim and Javaherian (1991) reported *Anopheles culicifacies* species from Iran. Nguen *et al.* (2000) described *Anopheles* (*Anopheles*) *nimpe* a new species in coastal area of Southern Vietnam. From Florida Darsie and Shroyer (2004) described *Culex* (Culex) *declarator* a new species. While, from oriental region Sallum *et al.* (2005) described six new species *of Anopheles leucosphyrus* group.

From India, Giles (1901a) described six new species of genus *Culex* and four new species of genus *Anopheles* (Giles 1901b). Similarly, Bentley (1902) described one new species of genus *Anopheles* from Tezpur district, Assam. Later, James and Liston (1904) published a monograph of the *Anopheles* mosquitoes of India. James and Liston (1910) also gave new classification to the genus *Anopheles*. During 1910 Theobald erected four new genera and described twenty new species of genus *Anopheles* and one new variety of *Anopheles* from India. James and Liston (1911) also published a monograph of *Anopheles* mosquitoes of India and enlisted thirty species of mosquitoes. While, from Simla Christophers (1911) described one new species of genus *Anopheles*.

Edwards (1923) described five new species of genus *Finlaya* from India likely Barraud (1923a) described five new species of subgenus *Stegomyia* from Assam. Barraud (1923b) also described two new species of *Culex* from Assam.

Barraud (1923c) presented a key to the Indian species of subgenus *Lutzia*.

In 1923 Senior White published a catalogue of Indian Culicidae. While, Barraud (1924a) have been described two new species of subgenus *Mochthogenes* from Kashmir. Barraud (1924b) also described four new species of subgenus *Finlaya* from the Western Himalaya one new species from subgenus *Lophoceratomyia* from Himalaya (Barraud, 1924c), two new species of genus *Culex* from Assam, three new species of genus *Culex* (Barraud 1924e), one new species of the subgenus *Culiciomyia* (Barraud 1924f), two new species of subgenus *Lophoceratomyia* (Barraud 1924g) from India and described five new species of genus *Culex* from Kashmir and North West frontier province (Barraud 1924h).

Christophers (1924) published a provisional list of mosquitoes and reference catalogue of the Anophelini in India. While, Barraud (1926) described five new species of *Uranotoenia* and *Harpagomyia* from India, two new species of genus *Aedomyia* and *Orthopodomyia* from Karwar (Barraud 1927a) and two new species from subgenus *Leicesteria* from Assam (Barraud 1927b).

Barraud (1927c) also described seven new species from subgenus *Aedimorphus* and *Finlaya* from India and one new species from subgenus *Mansonioides* from India (Barraud 1927d). Strickland and Chowdhury (1927) described one new species of Anopheline from Bengal. Barraud (1928) described eight new species of genus *Aedes* from India. From South India Puri (1929) described one new tree-hole breeding species of *Anopheles sintoni*. Barraud (1931a) described one species from subgenus *Stegomyia* from

Biharand eight new species of Indian Culicine mosquitoes (Barraud 1931b).

Barraud (1931c) described eight new species of subgenus *Aedes* two new species from genus *Stegomyia* from India Barraud (1931d) and Barraud (1932) described one new species from genus *Anopheles* from India. Christophers (1933) published key to the family Culicidae and tribe *Anophelini*. Similarly, Barraud (1934) also published key to the family Culicidae and tribes *Megarhinini* and *Culicini*. Later, Puri (1935) added key to the fully grown larvae of Indian *Anopheline* mosquitoes. Puri (1941) also published key to the Anopheline mosquitoes of India. While, Wattal and Kalra (1961) gave a region wise pictorial key to the female Indian *Anopheles*. Misra (1956) reported six species from Anopheline from Northeast frontier Agency. Reuben (1967) described one new species from subgenus *Diceromyia* from Southern India. Knight (1969) described one new species of the subgenus *Finlaya* from India. Later, Sen *et al.* (1973) described fourteen species of *Anopheles* from Tirap district Arunachal Pradesh. Hussainy (1981) described eighteen species of *Anopheles* and thirteen species of *Culex* from Bastar district of Madhya Pradesh. Reuben *et al.* (1994) published key to the species of *Culex* in Southeast Asia.

Reuben and Suguna (1983) provided morphological differences between sibling species of the *Anopheles subpictus* Grass from India. Likely, Rao (1984) published a key to the Anopheline mosquitoes of India. Tewari *et al.* (1987) reported thirty one species of *Anopheles*, eleven belongs to subgenus *Anopheles* and twenty belongs to subgenus *Cellia* from Western Ghats of Tamil Nadu. From Wynad district, Kerala, Tewari and Hiriyan (1990) described male larvae and pupae of *Aedes* (*Diceromyia) kanarensis* for the first time.

Rajput and Kulkarni (1990) reported thirty eight species and four varieties of *Culex* from Bastar district of Madhya Pradesh. They also reported nineteen species of *Anopheles* from Manipur. Rajput and Kulkarni (1991) described ten species from six genera of Bastar district of Madhya Pradesh. Tewari and Hiriyan (1991) described *Aedes (Rhinoskusea) protonovoensis* new species from South India. Pal and Dutta (1992) reported sixteen species of *Anopheles* from three district of Arunachal Pradesh. Rajput and Singh (1992a) reported twenty seven species of genus *Culex* from Manipur. Rajput and Singh (1992b) also gave distributional records of nineteen species of *Aedes* from Manipur. Rajput and Singh (1992c) also recorded eleven species from genus *Armigers* and single species from genus *Heizmannia* from Manipur. From South India Tewari and Hiriyan (1992) described two new species of subgenus *Diceromyia*. Tewari and Hiriyan (1994) described female pupa and larva of *Aedes (paraedes)* Barraud and the pupa and larva of *Aedes (paraedes) Menoni* for first time from Wynad district, Kerala. Rajavel (1996) reported *Culex (Lophoceraomyia) infantulus* Edwards from Pondicherry, South India. Bhattacharyya *et al.* (2000) reported *Armigeres joleoensis* Upper Assam. Rajavel *et al.* (2001) reported fifteen species of mosquitoes in Mangrove ecosystem of South India.

Recently, Sathe and Girhe (2001) studied biodiversity of mosquitoes in Kolhapur district, Maharashtra. Sathe and Girhe (2002) described *Aedes kolhapurensis, Aedes sangiti, Aedes punchgangi* and *Aedes indica* new species, they also studied seasonal abundance and distributional records of fifteen mosquitoes from Kolhapur district, Maharashtra. Bhattacharyya *et al.* (2003) described *Culex (Lophoceraomyia) quadripalpis, Culex (Lophoceraomyia) mammilifer* and

Uranotaenia (pseudoficalbia) novobscura as new species from Assam. Bhattacharyya *et al.* (2004b) also described *Verrallina (Neomacleaya) assamensis* and *Uranotaenia (pseudoficalbia) dibrugarhensis* from Assam. Rajavel *et al.* (2004) reported check-list of mosquitoes of Pondicherry. They reported sixty four species belonging to fourteen genera and twenty three subgenera. Sathe and Jagtap (2005) studied biodiversity of mosquitoes from Sangli city of Maharashtra.

Rajavel *et al.* (2005a) also reported forty three species of mosquitoes in mangrove forests of Bhitarkanik, Orissa. Rajavel *et al.* (2005b) reported *Culex (Lophoceraomyia) pilifermoralis, Culex (Lophoceraomyia) wilfredi* and *Heizmannia (Heizmannia)* chegi for the first time from Jaypore hills of Orissa. From Mandy district of Karnataka Satishkumar and Vijayan (2005) reported twenty nine species of mosquitoes in the rural areas.

Sathe and Tingare (2006) also studied biodiversity of mosquitoes in Solapur city, Maharashtra. Sathe and Tingare (2007) described one new species of the genus *Anopheles* Meigen from India. From Chorao mangrove of Goa Rajavel *et al.* (2007) reported fourteen species of mosquitoes. They also reported twelve species of mosquitoes in the mangroves of Vikhroli, Maharashtra. Bhattacharyya *et al.* (2007) reported *Topomyia (Topomyia) hirtusa, Topomyia (Topomyia) bifurcata* and *Topomyia (Suaymyia) cristata* for the first time from northeastern state of Arunachal Pradesh.

Very recently, Sathe and Jagtap (2008) studied the three decades trend of malaria from Sangli; they also studied the role of intensified mass surveillance campaign in Sangli district. Jagtap *et al.* (2009) reported the Chloroquine resistance in Etapalli, Gadchiroli. Sathe and Jagtap (2009) visualized the tree hole breeding and resting of mosquitoes in Western Ghats. Jagtap *et al.* (2009) reported the incidence of Dengue and shifting trend to rural in Kolhapur. Sathe and Tingare (2010) described the mosquito biodiversity of Southern Maharashtra, wherein they described 21 new species redescribed 8 species of mosquitoes. They also reported 48 species from Kolhapur region. Sathe and Jagtap (2010a) studied the biodiversity of Anopheline mosquitoes from Western Ghats and also the mosquito borne diseases (Sathe and Jagtap 2010b).

The review of literature clearly indicates that very little attention is paid on the taxonomy of mosquito from Maharashtra except the work of Sathe and his co-workers.

Seasonal Abundance

Review of literature indicates that the study of seasonal abundance is carried out by several workers. Senior White (1937) reported 23 *Anophelines* species on Jaypore Hills. 32 regions of mosquitoes in the world are visualized by Foot and Cook (1959). From Nainital Terai, Nagpal *et al.* (1983) reported seasonal abundance of 29 mosquitoes. Das *et al.* (1984) studied seasonal abundance of 42 species of mosquitoes from various places of Meghalaya. Rao (1984) reported 51 species of *Anopheles* and their subspecies and varieties from India. Nagpal and Sharma (1987) described seasonal prevalence of 61 species from Assam, Meghalaya, Arunachal Pradesh and Mizoram. Nagpal and Sharma (1995) found 14 species of the genus *Anopheles* from different part of India. From Southern India Reuben *et al.* (1992) reported seasonal prevalence of 10 species of *Culex* mosquitoes. Rajavel *et al.* (2000) studied seasonal prevalence of *Aedes portonovoensis* in mangrove forest of South India.

Sathe and Girhe (2001a) reported seasonal abundance of 4 *Anopheles* mosquitoes in Kolhapur region. Sathe and Girhe (2001b) studied seasonal abundance of 15 species of mosquito from Kolhapur district. Girhe and Sathe (2001c) visualized incidence of malaria in Kolhapur district. The seasonal prevalence of *Culex quinquefasciatus* is described in Godavari districts of Andhra Pradesh by Murty *et al.* (2002a). Murty *et al.* (2002b) studied a seasonal prevalence of *Culex vishnui* sub-group and *Anopheles* species in Andhra Pradesh during 1999. Kanojia *et al.* (2003) described seasonal prevalence of mosquitoes in Uttar Pradesh during 1990 to 1996. Sharma *et al.* (2005) reported seasonal abundance of *Aedes aegypti* in Delhi during 2003. In the irrigated and non-irrigated area of Thar, Rajasthan Joshi *et al.* (2005) reported seasonal abundance of *Anopheline* mosquitoes during August 2001 to July 2002. Tilak *et al.* (2006) attempted seasonal abundance of mosquito in Pune. In the Doon valley Dehradun Pemola and Jauhari (2006) reported seasonal abundance of 10 *Anopheles* and *Culicine* mosquitoes during 1999-2002. Malarial Research Centre (MRC) (2006) studied seasonal abundance and Bionomics of *Anopheles culicifacies, An. fluviatilis, An. minimus, An. sundaicus* and *An. stephensi* in India. Baruah *et al.* (2007) studied seasonal prevalence of malaria in Sonitpur district, Assam during 2002 to 2003. Recently, Sathe and Jagtap (2009, 2010) studied the tree hole breeding and resting of mosquitoes and the biodiversity of Anopheline mosquitoes in Western Ghats.

The review of literature indicates that little attention is paid on seasonal abundance of vector mosquitoes except the work mentioned above. The present work is precise attempt on the seasonal abundance of mosquitoes and will add great relevance in solving cases of mosquito borne diseases in the region.

Chapter 3
Materials and Methods

Correct identification of disease vector is of almost importance because the scientific name of an organism is the key to all known information about its morphology, its behavior and life history and its potential threat to human welfare. Vector species may be sampled to assess presence/absence or abundance in order to determine whether control measures are necessary. Following material and methods were adopted in completion of the work.

Materials

Mosquitoes resting on different surfaces (indoor or outdoor) were collected by hand collection method by using suction tube and insect net.

1) Suction Tube (Figure 4)

The suction tube contains a glass or plastic tube of 15 mm diameter and 20-30 cm length; a flexible rubber or plastic tube, 80-100 cm long and 15-18 mm in diameter.

The mesh between glass tube and rubber tube for preventing entry of mosquitoes and solid dusts towards the mouth. The third portion of the suction tube is mouthpiece for sucking mosquitoes by inhalating air through mouth. After sucking five mosquitoes, the sucked specimens were separated in the test tubes/specimen bottles.

2) Test Tubes (Figure 5)

Two types of test tubes one with open ends of both sides and another only one end open were used for mosquito transfer and storage (15 × 15 mm and 15 × 25 mm length and diameter). During the handling of adult mosquitoes the test tubes were plugged with cotton at the mouth or both the ends in case of open tubes.

3) Torch (Figure 6)

The Torch is essential for the adult surveillance identification of resting mosquitoes in various habitats in day or night.

4) Mosquito Rearing Cage (Figure 7)

Mosquito larvae collected from field spots in beakers were reared in the cage (size of the cages was 25 × 25 × 25 cm). Three sides contain nylon mesh and at one side, with muslin cloth, sleeve as door for easy handling of mosquitoes.

5) Specimen Tubes (Figure 8)

Specimen tubes of size, 6 × 2 cm, 5 × 2 cm and 4 × 2 cm (length and diameter) were used for preserving and handling the adult mosquitoes. The open end of specimen tube was plugged either by rubber or wooden cork/cotton balls. Specimen tubes were used for keeping pinned adult mosquitoes.

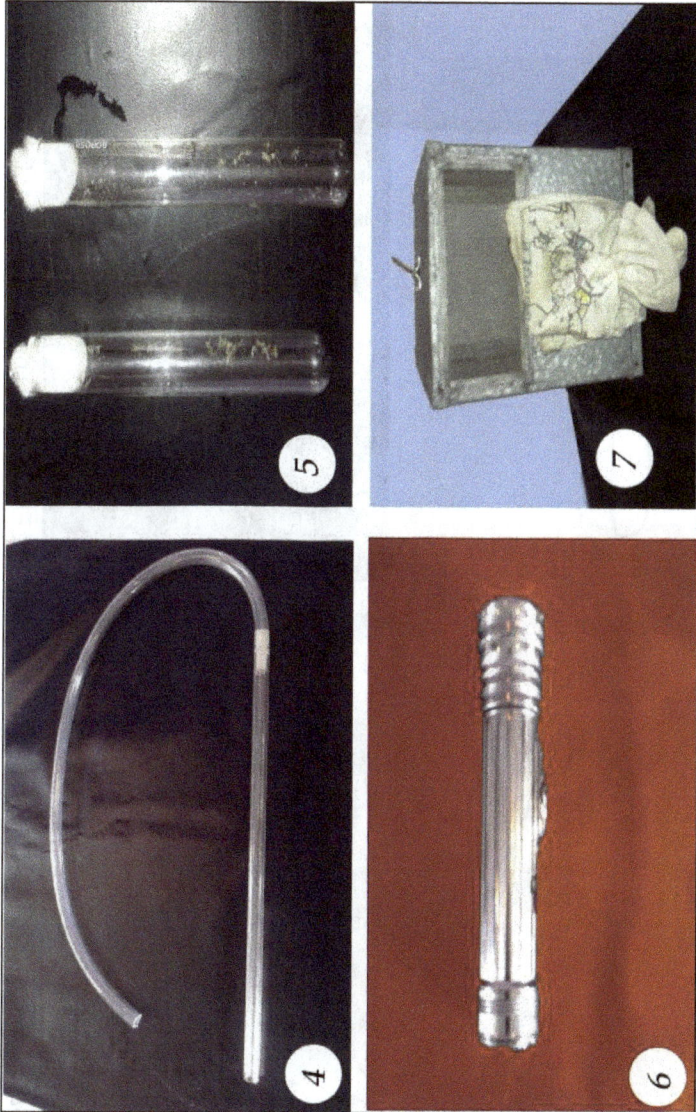

Plate 1–Figure 4: Suction Tube; **Figure 5**: Test Tube; **Figure 6**: Torch; **Figure 7**: Mosquito Rearing Cage

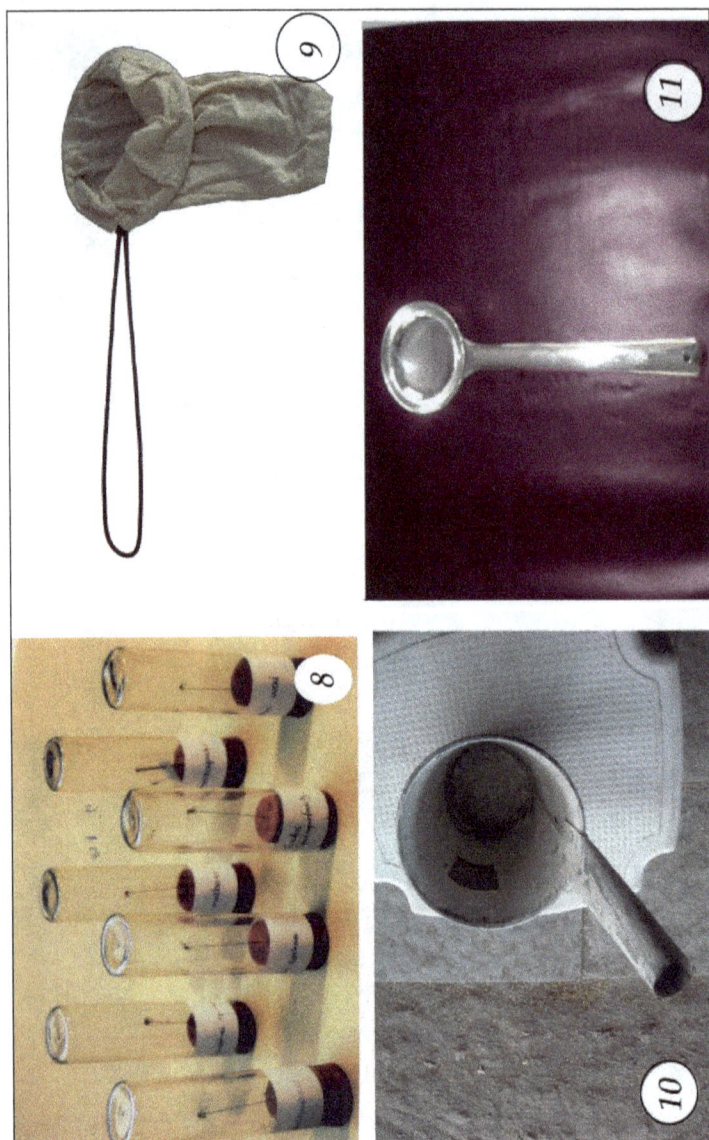

Plate 2–Figure 8: Specimen Tubes; **Figure 9:** Insect Net; **Figure 10:** Dipper; **Figure 11:** Ladle

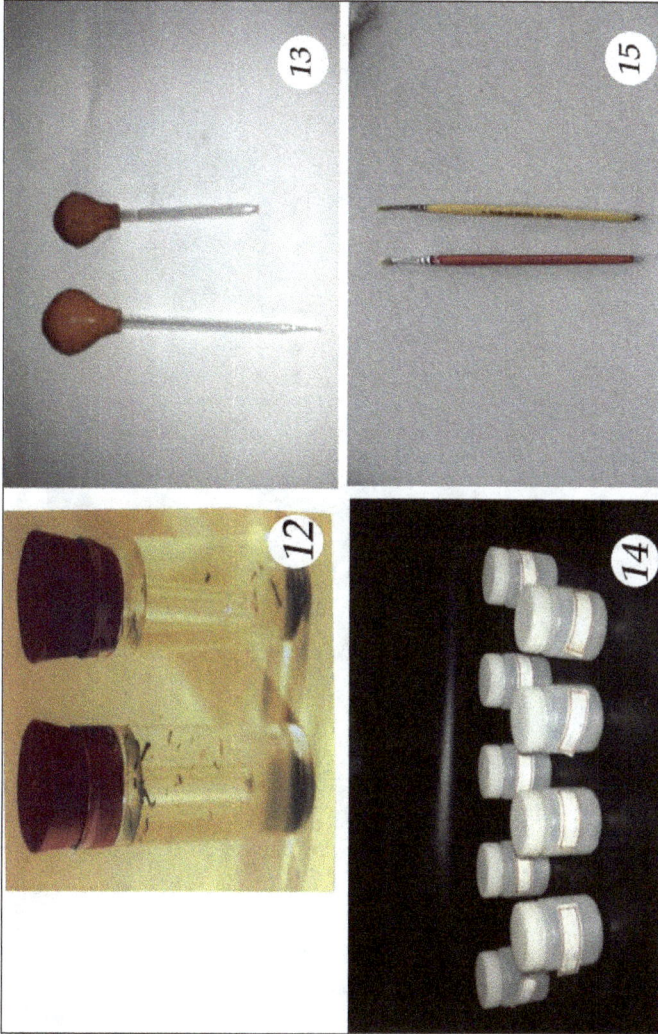

Plate 3–Figure 12: Specimen Bottles; **Figure 13**: Dropper; **Figure 14**: Plastic Containers; **Figure 15**: Camel Brush.

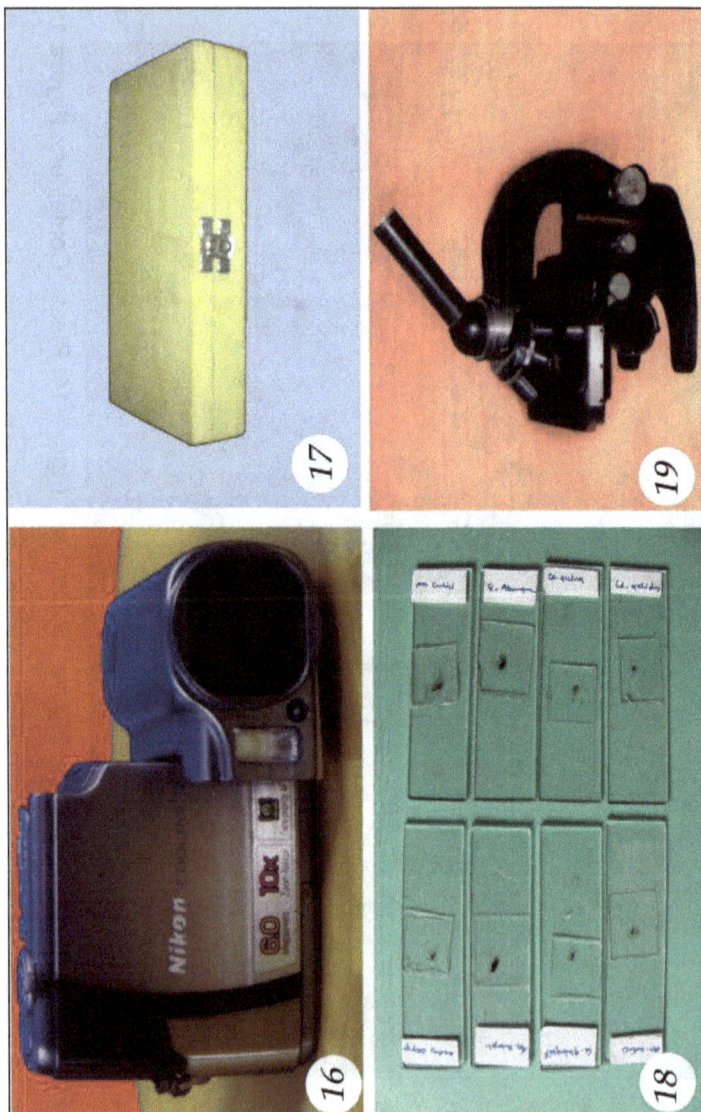

Plate 4–Figure 16: Camera; **Figure 17:** Slide Box; **Figure 18:** Slides and Coverslips; **Figure 19:** Microscope

6) Insect Net (Figure 9)

Outdoor and indoor mosquitoes were collected by hand net. Insect hand net contain aluminium handle 70 cm long, circular iron ring of 22 cm. diameter and ordinary mosquito nylon mesh bag of 70 cm depth.

7) Dipper (Figure 10)

The dipper (Figure 10) was used for collecting and handling mosquito larva and pupae from natural habitat. The larvae and pupae will remain in the dipper and the water will be dropped down.

8) Ladle (Figure 11)

Ladle is used for collecting mosquito larvae and pupae from various aquatic habitats of mosquitoes. It consists of a flat handle of length 25 cm with a circular concave cap of diameter 8 cm (Figure 11).

9) Plastic Containers (Figure 12)

The larvae and pupae collected must arrive alive and undamaged at the laboratory. Hence, plastic containers were used for collecting the samples of larvae/pupae from the natural habitat.

10) Rubber Droppers (Figure 13)

Rubber droppers were used for picking the larvae/ pupae from natural habitat and handling them in laboratory.

11) Specimen bottles (Figure 14)

The adults and larvae were preserved in specimen bottles (size, 6 × 2 cm, 5 × 2 cm and 4 × 2 cm length and diameter). The mouth of bottle was closed by rubber cork/ plastic cap for keeping specimen air tight and safety.

12) Camel Hair Brushes (Figure 15)

Camel hair brush No. 1 and 2 has been used for handling the specimen and preparation of slides.

13) Camera Nikon S4 (Figure 16)

Nikon Coolpix S4 camera, 6 megapixel and 10x optical zoom was used for photography of the mosquitoes and mosquito habitat.

14) Slide Box (Figure 17)

Slide box were used for keeping the permanent slides safely. Slide boxes of size, 28 × 22 × 3.5 cm, 21 × 19 × 3.5 cm (length, width and height) were used.

15) Slides and Cover Slips (Figure 18)

Ordinary slides and cover slips were used for preparing the whole mounts and other body parts such as head, proboscis, antenna, thorax, wing, halter, hind leg and abdomen.

16) Oven

Oven of size, 3.6 × 2.4 feet (height and width) has been used for drying adult mosquitoes and the permanent slides of mosquitoes.

17) Compound Microscope (Figure 19)

Simple monocular compound microscope with objectives 10x, 45x, 100x were used for describing the mosquito species with the help of occulometer.

18) Chemicals

Following chemicals were used for preparation of slides and preserving the insects.

    1. 10 per cent KOH.

2. 30 per cent to 100 per cent Ethyl alcohol grades.

3. Glacial acetic acid etc.

4. Xylene.

5. DPX/Canada Balsum.

Methods

The survey of mosquitoes was made from Western Maharashtra (districts Pune, Satara, Kolhapur and Sangli) (Figures 2 and 3) from 2005 to 2011. A large number of specimen were collected by visiting various places of Western Maharashtra namely, Pune (Baramati, Pune, Bhor, Saswad, Haveli, Junner), Satara (Medha, Wai, Mahabaleshwar, Satara, Patan, Mhaswad, Koregaon), Kolhapur (Ajra, Malkapur, Kagal, Kolhapur and Jaysingpur) and Sangli (Miraj, Vita, Tasgaon, Shirala and Jath) at 15 days interval, by one man one hour search method.

Most commonly, a mixture of 75 per cent alcohol to 25 per cent water was used for preservation of insects. The water should be distilled to ensure a neutral PH and the solution should be thoroughly mixed since alcohols and water do not mix easily by themselves. The adult mosquitoes were narcotized in ether and killed in killing bottle by chloroform. Specimen for molecular work should be collected in 95 per cent or absolute (100 per cent) ethanol (ethyl alcohol).

The mosquitoes were pinned with entomological pins from the ventral side, kept on spreading board and dried in drying chamber/oven at 60°C. The dried specimen then kept in Sterilized specimen tube by pinning invertedly to wooden cork. Pinning refers to the insertion of a standard insect pin directly through the body of an insect from ventral side using care that the pin does not tear off any legs or destroy chetotaxy on the dorsal side of the insect. After the pin is inserted and before the specimen is dry, the legs, wings, and antennae should be arranged so that all parts are visible for study. Pinned specimens should always be placed as in a small box with a foam pinning bottom. The box should be well wrapped and placed in a larger carton.

Specimen are mounted so that they may be handled and examined with the greatest convenience and with the least possible damage. Well-mounted specimen enhance the value of a collection; their value for research may depend to a great extent on how well they are prepared and preserved. For the taxonomical study, head, antenna, proboscis, wing, legs, halter and abdomen were mounted on slide in D.P.X. Specimen have been prepared on the slide and the slides are kept in slide boxes. The final stage in preparing permanent mounts is through drying or hardening of the medium. This may be done in any clean environment or in an oven or special slide warmer under gentle heat. The mounts should be carefully labeled either before drying or afterward. The records were made on locality, date of collection and identification. To label microscope slides, square labels are used. Morphological studies were carried out with the help of monocular microscope. Comparative measurements of body parts of specimen were made with ocular micrometer and calculated with the help of graduated mechanical stage. All measurements were made in millimeter.

The terminology adopted for description of species is same as that of Christophers (1933), Barraud (1934), Puri (1948), Horsefall (1955), Knight and Stone (1977), Tanaka

(1979), Rao (1984), Nagpal and Sharma (1995), Sathe and Girhe (2002) and Sathe and Tingare (2010). A large number of references were consulted during the course of the study and cited in the text.

Morphological Consideration

Adult (Figure 20) mosquitoes are slender bodied and smaller insects and posses forwardly projecting proboscis, numerous scales on the thorax, legs, abdomen and wing veins; a fringe of scales on the wings, only one pair of functional wings, pair of small knob-like halters.

Head (Figures 21 to 22)

The features of head are shown in (Figures 21 and 22). The anterior section of the body bears the compound eyes, antennae and mouth parts. In females, long, slender, scaled proboscis present. Males of most mosquitoes can be distinguished from females by their verticillate (bushy, not plumose) antennae.

Antenna (Figures 20 and 22)

A pair of filamentous and segmented antennae is situated in between the eyes. In females the antennae are pilose having short hairs, but in males (Figure 22) they are feathery or plumose. Each antenna is divided into three segments the basal segment, the scape, second segment pedicel and the third segment is flagellum, which is sub-divided into 13 or 14 segments called flagellomeres.

The Behind the compound eyes on dorsal side (Figure 21) consists of the vertex and occiput and is more or less covered by scales. The proboscis (Figure 22) is covered with scales and terminates in a pair of lobe-like labella.

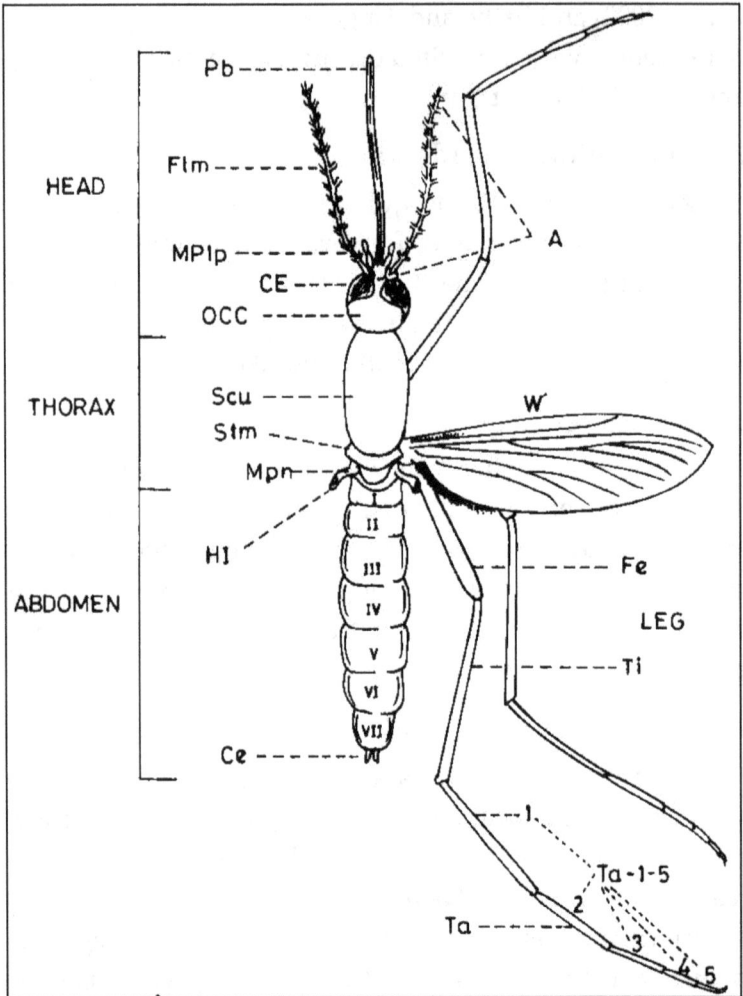

Figure 20: Dorsal View
Pb: Proboscis; Film: Flagellomere; MPlp: Maxillary Palpus;
CE: Compound eye; Occ: Occiput; Ap: antepeontum; Scu: Scutum;
Stm: scutellum; H1: Halter; Mpn: Mesopostnotum; Ce: Cersus;
A: antenna; W: wing; Fe: Femur; Ti: Tibia; Ta: Tarsus; Ta-1-5:
Tarsomeres.Occiput; Pb: Proboscis; Plp: Plapus.

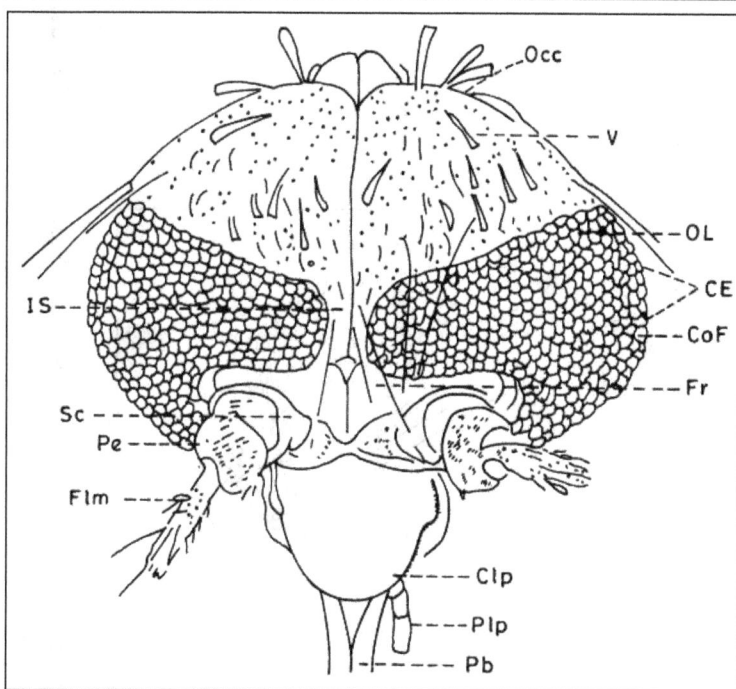

Figre 21: Female Head

IS: Interoccular space; Sc: Scape; Pe: Pedicel; Film: Flagellomere;
Clp: Clypeus; Fr: Frons; CoF: Corneal Facet; CE: Compound eye;
OL: Occular Line; V: Vertex; Occ: Occiput; Pb: Proboscis;
Plp: Plapus

Thorax (Figures 23 and 24)

Three segmented thorax bears a pair of legs in each segment. The largest or middle segment of the thorax bears the wings and third segment bears a pair of the halters. Behind the scutum is the crescent-shaped scutellum may be covered with scales and setae along the posterior edge.

Legs (Figure 25)

Long and slender legs are covered with scales, usually brown, black or white in color. Each leg is divided by six

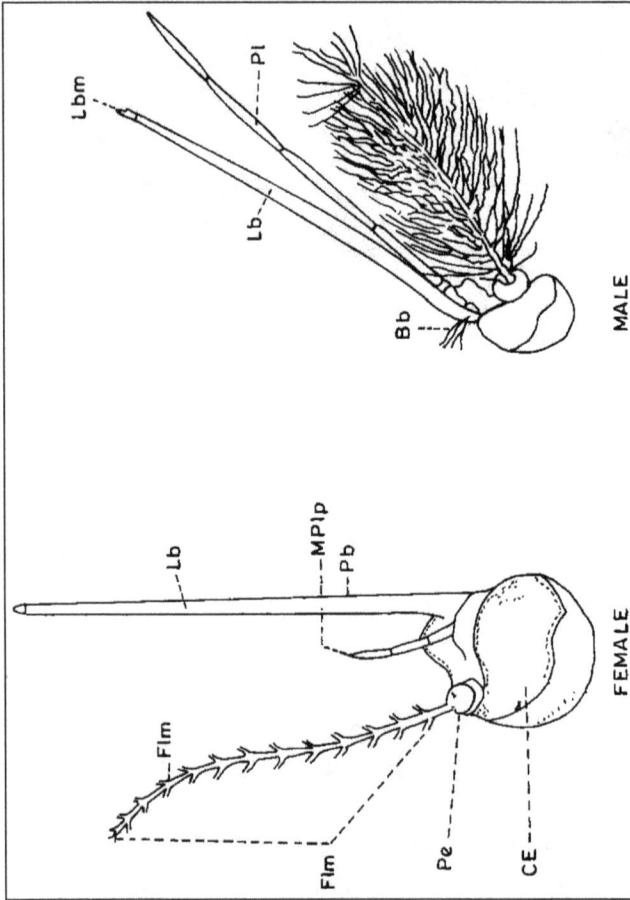

Figure 22: Male and Female Head

Flm: Flagellomere; Flm: Flagellum; Pe: Pedicel; CE: Compound Eye; Pb: Proboscis; MPlp: Maxillary Palpus; Lb: Labium; Pl: Palpus; Lbm: Labillum; Bb: Bosal Bristles

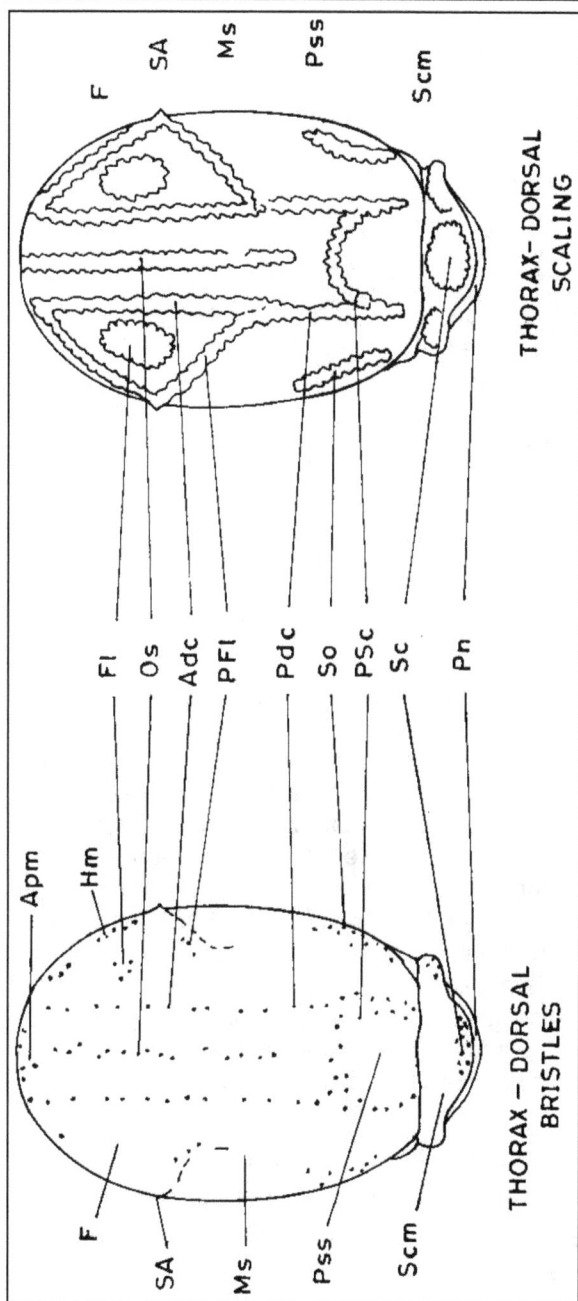

Figure 23: Thorax View

Apm: Anterior Promantory; Hm; Humerol; FL: Fossal; Os: Ocrostichol; Adc: Anterior-dorsocentrol; So: Supraolar; PFl: Posterior Fossal; Pdc: Posterior Dorsocentral; Psc: Prescitellar; Sc: Scutellar; Scm: Acutellum; Pn: Postnotum; F: Fossa; Ms: Mesonotum; SA: Scutal; Pss: Pre Scutellor Space.

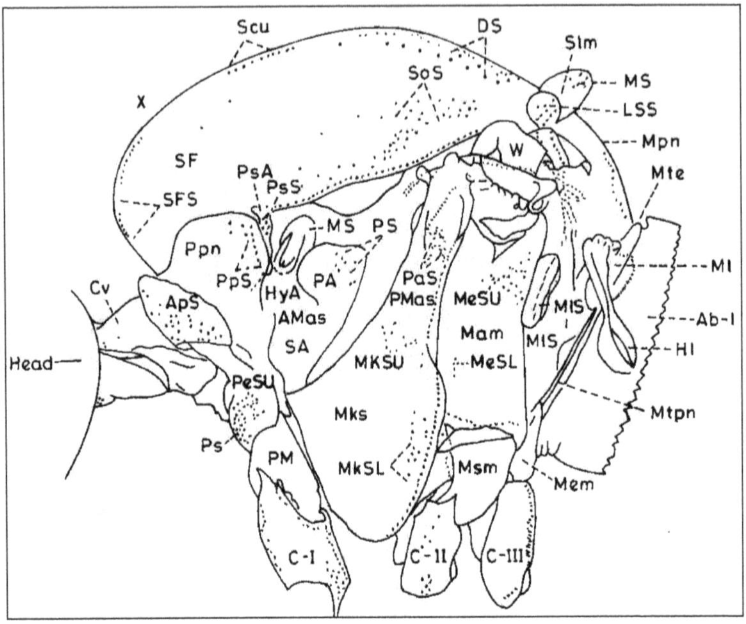

Figure 24: Thorax View

Cv : Cervix; DS: Dorsocentral Setae; H1: Halter; PM: Post Procoxal Membrane; Ppn: Postpronotum; Pps: Post Pronotal Setae; PS: Postspiracular Setae; Ps: Proepisternum; Ab-I: Abdominal Segment I; Amas: Anterio Mesanepisternum; SF: Scutal fossa; SFS: Scutal Fossal Setae; Stm: Scutum; HyA: Hypostigmal Area; LSS: Lateral Scutellar Setae; Mam: Mesanepimeron; Mem: Metameron; MeSL: Lower Mesanepimeral Setae; MeSU: Upper Mesanepimeral Setae; PsA: Prespiracular Area; PsS: Prespiracular Setae; Scu: Scutum; SA: Subspiracular; SoS: Supralar Setae; Mks: Mesokatepisternum; MkSL: Lower Mesokatepisternal Setae; MkSU: Upper Mesokatepisternal Setae; Mpn: Mesopostnotum; MS: Mesothoracic Spiracle; Msm: Mesomeron; Mtpn: Metathoracic Apiracle; PA: Postspiracular Area; PaS: Prealar Setae.

segments, the coxa, trochanter, femur, tibia, tarsus and posttarsus. The tarsus consists of five segments called tarsomeres. The tarsus usually may be in a pair of toothed or simple claws.

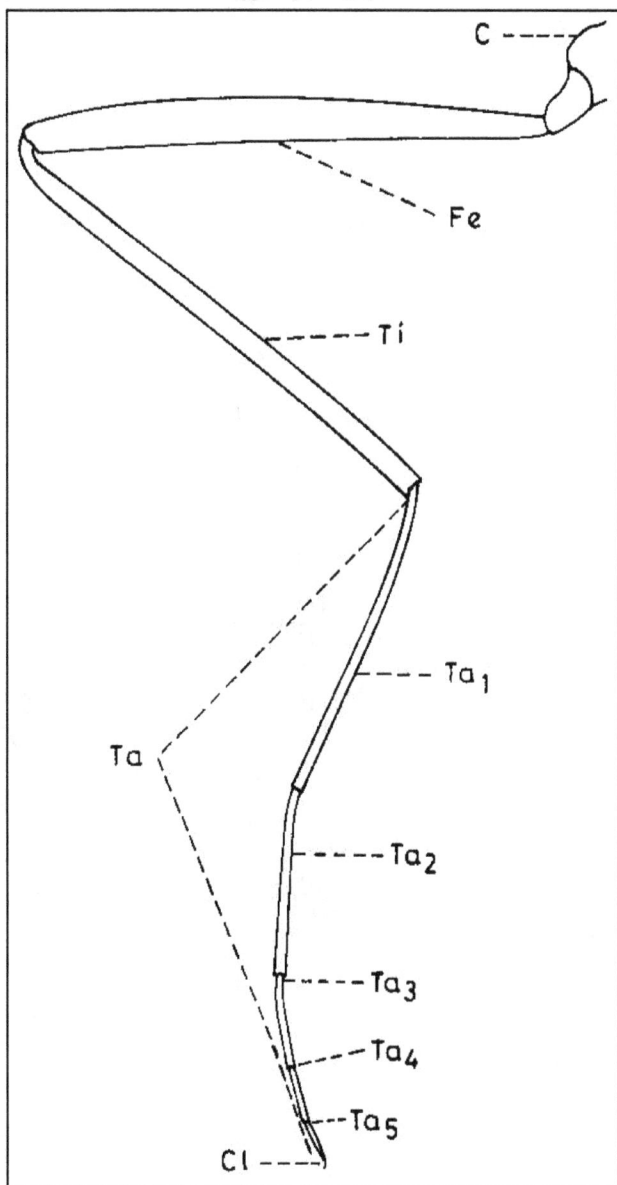

Figure 25: Leg
C: Coxa; Fe: Femur; Ti: Tibia; Ta: Tarsus; Ta: 1 : 5 : Tarsomere;
Cl: Claw.

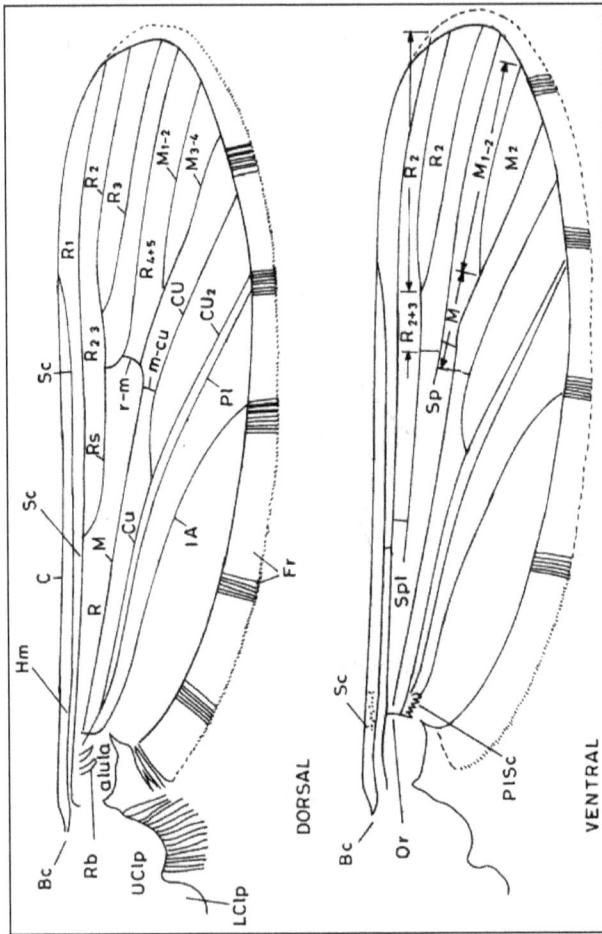

Figure 26: Wing

C: Costa; R: Radius; Sc: Subcosta; M: Media; Rs: Radial Sector; 1A: First Anal vein; Bc: Basicosta; Hm: Humaral; Sc: Subcosta; Fr: Fringe; Pl: Plical; Rb: Remigial Bristles; LCip: Lower Calypter; UCip: Upper Calypter; Sp: Spur; Or: orculus; PLSc: Plical Scales

Wings (Figure 26)

The wings are long and relatively narrow and the veins are covered with brown, black, white or creamy yellow scales. The principal longitudinal veins of mosquitoes are the costa, subcosta, radius, media, cubitus and anal vein. The veins are clothed with scales and posterior margin of the wing bear a fringe of scales (wing fringe).

Abdomen (Figure 20)

The abdomen is the third, posterior division of the body which contains 10 segments but only the first seven or eight are visible. Segments IX and X are reduced and more or less modified in genitalia.

Photography

The pinned mosquitoes and whole mounts of the mosquitoes were considered for microscopic photography. The colored photography of whole mounts was made with the help of Nikon S4 camera (6.2 megapixels, 10 X optical zoom). Whole mount figures have been enlarged 16 times to its original size.

Chapter 4
Taxonomy

Taxonomy plays a crucial role in the field of basic sciences and applied sciences such as pest management, forestry, environmental problems, wild life management, nutritional science, forensic science, public health, national defense, biotechnology, nanotechnology and several other fields in identifying the species. Taxonomy, biology and applied sciences are interlinked. Taxonomy helps in understanding evolutionary history of the species and guides the explorations for native and exotic species for their use in sustainable development of region or country.

Perusal of literature indicates that James and Listone (1911), Knowles and White (1927), Strickland and Choudhury (1927), Korke (1928, 1932), Barraud (1934), Puri (1947, 1948, 1960), Basu (1958), Bhatia *et al.* (1958), Stone *et al.* (1959), Christophers (1960), Reid and Knight (1961), Wattal and Karla (1961), Smart (1965), Reid (1966), Roy and Brown (1970), Harrison *et al.* (1973), Rao (1975, 1984),

Knight and Stone (1977), Chamnarn (1986), Subharao *et al.* (1988), Das *et al.* (1984, 1990), Nair and Mathew (1993), Nagpal and Sharma (1987, 1995), Sathe (2006), Sathe and Girhe (2001, 2002), Sathe and Jagtap (2005), Sathe and Tingare (2006, 2010) etc. have been worked on taxonomy of Indian mosquitoes while, Christophers (1933), Barraud (1934), Edwards (1941), LaCasse and Yamaguti (1950), Mattingly (1952, 1965, 1969), Lane (1953), Foote (1954), Carpenter and Casse (1955), Vansomeren *et al.* (1955), Laird (1956), Foote and Cook (1959), Stone *et al.* (1969), Steward and McWade (1961), Stone (1961, 1963, 1967, 1970), Belkin (1962, 1970), Mithalyi (1963), Vanden Assem and Bonne (1964), Dobrotworsky (1965), Smart (1965), Carpenter (1966), Delfinado (1966), Taylor (1967), Gillies and Meillon (1968), Zavortink (1968, 1970) Reid (1968), Smith (1969), Gutsevich *et al.* (1970), etc. worked on mosquito taxonomy from different parts of the world. Additional workers on taxonomy of mosquitoes are included in bibliography.

The mosquitoes are grouped under Diptera. The Diptera is the fourth largest order in class insecta and subdivided into sub orders Nematocera, Brachycera, Aschiza and Schizophora. The mosquitoes belong to Phylum Arthropoda, class Insecta, suborder Nematocera and family Culicidae.

The Order Diptera has been classified by Mc Alpine (1979). He arranged Diptera into in two suborders Nematocera and Brachycera, while Mani (1982) divided into 4 sub orders namely Nematocera, Brachycera, Aschiza and Schizophora. The family Culicidae is grouped under Nematocera in adition to 13 families. The Culicidae shows following features:

1. Elongated and piercing and sucking type of mouth parts, males are cell sap suckers.

2. Long and filamentous antenna with 14 or 15 segments, brushy and plumose antenna in male and hairy and pilose in female.

3. In between prescutum and scutum a definite suture absent, pronotum completely divided.

4. Scales present on wing, head and body.

5. No cross-vein connection of R_1 and R_2, media (M) two branched and anal vein (A) long and reaching wing margin, vein subcosta (SC) reaching the costa, R_{2+3} forked, R_{4+5} simple.

6. A pair of functional spiracles present dorsally on 8th abdominal segment, larvae aquatic, with complete head capsule.

The family Culicidae is further divided into three sub families namely, *Culicinae*, *Anophelinae* and *Toxorhynchitinae*. The Sub family *Culicinae* contain thirty genera, the Sub family *Anophelinae* contain three genera while, the Sub-family *Toxorhynchitinae* contain only one genus. The subfamily *Culicinae* shows following characters:

1. Trilobed scutellum with bristles and area between the lobe without bristles.

2. Palpi shorter than the proboscis in female.

3. Palpi long and with uniform thickness throughout the entire length observed in male, terminal part generally with little bend.

4. Scales are present on veins as well as along hind margin of wing.

5. Abdomen entirely covered with broad scales which always lie flat.

6. Piercing and long mouth parts in female.

The subfamily *Culicinae* includes about 1500 species belonging to more than 20 genera. Two third of the total described species of mosquitoes are scattered under the genera *Culex* and *Aedes*.

Subfamily *Anophelinae* is further subdivided into three genera *viz. Anopheles, Chagasia* and *Bironella*. Some members of genus *Anopheles* are the vectors of human malaria throughout the world. However, the last two genera are not involved in the transmission of malaria and do not reported from Indian region. The subfamily *Anophelinae* is visualized by following features:

1. Rounded scutellum.

2. Larvae predaceous.

3. Pulvilli and tibial bristles absent.

4. Scales absent on abdomen.

5. In both male and female palpi as long as proboscis.

6. Long and slender legs.

Key to the Genera of Tribe Anophelini

1. In male 2 large claws on II and III legs, Slightly trilobed scutellum, with a set of Bristles on each lobe larvae bearing a fringe of hairs .... *Chagasia*

2. In male single claw on foreleg and without Median or basal spur, scutellum Bar shaped Or evenly rounded with continuous line

of bristles, Vein 5.1 concave beyond cross
vein, larvae without fringe of hairs .... *Bironella*

3. In male single claw on foreleg and with
median or basal spur except in, *An. culiciformis*
which has no basal spur scutellum Bar shaped
or evenly rounded with continuous line of
bristles, Vein 5.1 not concave beyond cross
vein, larvae without fringe of hairs .... *Anopheles*

Genus *Anopheles* Meigen 1818

Meigen 1818: 10. Evans 1938: 1 (Ethiopian Region). De
Meillon 1947: 1 (Ethiopian Region). Jepson, Moutia and
Courtois 1947 : 177 (Mauritius). Collesss 1948 : 71 (Borneo).
Chow 1949 : 121 (China). Doucet 1951 : 1 (Madagascar).
Romeo viamonte and Castro 1951 : 313 (♀ buccopharygeal
armature). Lane 1953 : 186, (Neotropical Region), Pringle
1954 : 53 (Iraq). Weyer 1954 : 3 (Palaegrctic Region). Gelfand
1954 : 1 (Liberia).

Meigen. syst. Besshr. I.P. 10, (*Anopheles* Hgg.). The
genus *Anopheles* is errected by Meigen in 1818. It shows
following features:

1. Club shaped palpi in male.

2. Pronotum lack setae.

3. Five segmented palpi.

4. Palpi as long as proboscis and slender in female.

5. Half moon shaped or rounded scutellum with a
uniform row of hairs along the margin.

6. Proboscis in straight line with body.

7. Mandibles and maxillae of female well developed
and toothed.

8. At rest body not hump backed.

9. Wing spotted with white and dark scales are present.

10. Densely hairy abdomen.

11. Very small anal cerci.

The genus *Anopheles* contain more than 420 species from world. Out of which 50 are well known vectors of malaria. Christophers (1933) divided the genus *Anopheles* into four subgenera.

1. *Anopheles*
2. *Myzomyia*
3. *Nyssorhynchus* and
4. *Stegomyia*.

Nagpal and Sharma (1995) visualized six subgenera namely:

1. *Anopheles* Meigen,
2. *Cellia* Theobald,
3. *Kerteszia* Theobald,
4. *Lophopodomyia* Antunnes,
5. *Nyssorhynchus* Blanchard and
6. *Stethomyia* Theobald of this genus.

Twelve zones of epidemiology were observed by Nagpal and Sharma (1995) of the world. The subgenus *Anopheles* found in all the 12 zones with a maximum distribution in malaysian (56 species) and minimum from Afrotrophical zones (11 species). The subgenus *Cellia* is found in 10 zones and abundant in Afrotrophical zone (115 spp.) and a minimum from Southern American zone (1 sp. only). Three American zones are covered by subgenus *Kerteszia*, with a maximum number from South America (11 spp.), followed by Central America (5 spp.) and North America (1 sp. only). The subgenus *Lophopodomyia* is reported from two zones, maximum from South America (16 spp.) and Central America (1 sp. only). The subgenus *Nyssorhynchus* is reported from only four epidemiological zones with a maximum distribution in South America (30 spp.) followed by Central America (9 spp.). The subgenus *Stethomyia* is reported from two regions only, South America (5 spp.) and Central America (2 spp.). 197 species have been reported from the world and 34 species from India from the subgenus *Cellia*, (Nagpal and Sharma, 1995).

Recently, 58 species of the genus *Anopheles* have been reported from India (Rao, 1984; Nagpal and Sharma, 1995; Sathe and Girhe, 2002). Sathe and Girhe (2001, 2002) reported four species of *Anopheles* have been reported from Kolhapur district, Maharashtra.

Key to the Subgenera of the Genus *Anopheles*

1. Wings completely dark or if pale areas present, dark areas on costa involving both costa and vein 1 are less than four in number; parabasal spines of male two; pharyngeal armature absent .... Subgenus *Anopheles*

2. Wings always with pale and dark markings; dark areas on costa involving also vein 1 four or more in number; parabasal spines of male four or five in a group .... Subgenus *Cellia*

Key to the Species of Subgenus *Cellia*

1. Tip of hind tarsi not white 2

   Tip of hind tarsi white 17

2. Tarsi of front legs with broad pale bands 3

   Tarsi of front legs unbanded or only with
   narrow bands 6

3. Femora and tibiae speckled 4

   Femora and tibiae not speckled. 5

4. Female palpi with both apical and preapical
   pale bands broad and one narrow more basal
   band, sometimes speckled; thorax with broad
   scales .... *stephensi*

   Female palpi with only the apical pale band
   broad; preapical band narrow; thorax not
   covered with broad scales .... *sundaicus*

5. Palpi of female with dark .... *subpictus* and
   preapical area equal to or nearly var .... *vadakadiensis*
   equal to the pale apical bands*
   Palpi of female with dark preapical
   area half or less than half the length of .... *vagus*
   the apical broad band*

6. Thorax with obvious scales 7
   Thorax with hairs or hair-like 11
   scales only

7. Tip of female palpi dark 8

   Tip of female palpi not dark 9

8. Fossae in both sexes covered .... *multicolor*
   with scales. Fossae devoid of scales .... *turkhudi*

9. Tarsi with narrow but distinct 10
pale bands

Tarsi bands absent or indistinct .... *superpictus*
and not white .... (not recorded in India)

10. A line of overlapping broad white .... *moghulensis*
scales on side of thorax in front of...............................
wing roots

Without such a line of scales, ... *jeyporiensis*
scaling confined to median area .... and var.
of dorsum of thorax...................................... *candidiensis*

* At present males are indistinguishable.

11. Spotting of wings confined to .... *dthali*
costa and vein 1 only, rest of the .... (Kashmir only)
wing dark; head scales narrow.............................
and rodlike.

Ratio of femur and 1st tarsal 2:1,
White scales are present on head .... *mahabaleshwari*
.... sp. nov.

Wing with usual wing spots on all............................ 12
veins; head scales of ordinary type.

12. Female palpi with two broad pale............................ 13
apical bands, as broad or broader
than the intervening dark area Female
palpi with subapical pale 15
band narrow; intervening area
very broad.

13. Fringe spot present on vein 6 : ....*aconitus*
apical half of proboscis pale.
No fringes spot on vein 6; 14

proboscis dark or apical half
pale in certain lights only.

14. Basal third of cost uninterruptedly
    dark, without pale interruption, ... *varuna*
    not even with a pale scale; outer
    half of proboscis faintly or more
    markedly pale in certain lights.
    Basal third of costa with a pale .... *minimus*
    interruption however small;
    proboscis with apical half dark;
    except sometimes with a pale
    spot ventrally.

15. Fringe spots well marked at all 16
    veins except 6; some erect pale
    scales in front of thorax; vein
    1 pale in basal area.
    Fringe spot on two veins only
    except rarely; no pale scales or
    very few in front of thorax; vein
    1 internal to inner dark costal
    spot with dark spot (Vein 1 with
    a dark spot opposite the pale
    interruption on costa outside
    humeral cross vein); third
    vein mostly dark. .... *culicifacies*

    Palpi is not equal to proboscis,
    Fringe spot on vein 4.1, 5.2 and 6 .... *karveeri* sp. nov.

16. Third vein usually extremely
    pale; thorax with median area
    markedly paler than dark sides;
    frontal tuft conspicuous (Inner

quarter of costa entirely dark).........................*fluviatilis*
Third vein all dark or with only
a pale spot; thorax uniformly
coloured; frontal tuft poorly .... *sergentii*
developed. .... (Not found in India)

17. Hind tarsi with only one segment
    or less white, commonly with 18
    white bands above this.
    Hind tarsi with a continuous
    white area embracing atleast two 23
    terminal segments.

18. Femora and tibiae not speckled. 19
    Femora and tibiae speckled. 20

19. Female palpi with two broad pale
    apical bands and one narrow
    band near base in addition to
    usual more basal band .... *karwari*
    (Total 4 bands)

    Female palpi with two broad
    pale apical bands and the
    usual basal band only .... *majidi*
    (Total 3 bands)

20. Sixth vein with not more than 21
    three dark spots.
    Sixth vein with more than three
    dark spots. 22

21. Abdomen with a row of
    continuous black scale tufts on
    ventral surface of all segments
    easily visible in the lateral view

to naked eyes; female palpi
with four pale bands .... *kochi*
Abdomen not so, female palpi .... *maculatus* and
with usual three bands only. .... var. *willmorei*
Speckling on legs, white scales .... *krishnai* sp. nov.
on head

22. Tibio-tarsal joint of hind leg .... *balabacensis*
    with a broad conspicuous .... and *elegans*
    white band.
    Tibio-tarsal joint of hind leg .... *tessellatus*
    without a broad band.
    4 pale bands on palpi, preapical ....
    Pale band short......... .... *waii* sp. nov.

23. Femora and tibiae not speckled. 24
    Femora and tibiae speckled. 26

24. Hind tarsi 3¾ segments
    continuously white; abdomen
    heavily clothed with broad
    scales, which form lateral tufts .... *pulcherrimus*
    except on the last few segments.
    Hind tarsi 3 ½ or less segments
    continuously white; abdomen in
    with atmost rather narrow scales
    forming tufts except on last few
    segments 25

25. Vein 5 mainly dark; or with
    atleast a dark spot about the
    middle near the origin of the
    branch .... *annularis*

Vein 5 continously pale except .... *pallidus,*
at base and apex. .... *philippinensis nivipes*
.... (not in India)

26. Hind tarsi with only two .... *theobaldi*
    segments completely white.
    Hind tarsi with three segments
    completely white. 27

27. Female palpi with two broad
    apical pale bands and a narrow
    band and conspicous speckling;
    male palpi with shaft banded
    and spotted with white. .... *splendidus*
    Female palpi with broad apical
    pale bands or two narrow pale
    bands, without speckling;
    male palpi with shaft dark. 28

28. Dorsum of last two abdominal
    segments clothed with golden hairs
    and scales; inner quarter and
    outer third of costa chiefly pale .... *jamesii*
    Dorsum of last two abdominal
    segments not so, inner quarter
    of costa and outer third of
    costa chiefly dark. .... *ramsayi*

Anopheles (Cellia) *culicifacies* Giles, 1901

Female (Figure 55)

3.95 mm long, 0.70 mm broad; head 0.75 mm long,
0.60 mm broad, blackish brown; antenna 1.85 mm long,
brownish; thorax 1.50 mm long, 0.65 broad, blackish brown;
forewing 3.45 mm long, 0.70 mm broad, yellowish; hind

leg 6.55 mm long, yellowish; abdomen 1.75 mm long, 0.90 mm broad, reddish brown.

Head (Figure 27)

0.75 mm long, 0.60 mm broad, blackish brown, globular, with narrow scales; compound eyes black, rounded, ocular space 0.25 mm, interocular distance 0.19 mm long; vertex smooth, dark brown; nape 0.15 mm long, 0.17 mm broad, tubular, brownish; proboscis 2.20 mm long, blackish, scally, cylindrical; labium 0.25 mm long, brown, yellowish, scally; labellum 1.95 mm long, brownish, scally; palpi 2.20 mm long, 5 segmented, slender, as long as proboscis, densely scaly. Apical pale band nearly equal to the pre-apical dark band; mandibles and maxillae well developed, yellow.

Antenna (Figure 28)

1.85 mm long, 15 segmented, hairy, brownish, pilose; scape 0.05 mm long, 0.12 mm broad, brownish; pedicel 0.27 mm long, 0.11 mm broad, yellowish brown; flagellum 1.53 mm long, 13 segmented.

Flagellar Formula

3 L/W = 3.33, 9 L/W = 3.5, L 3/9 = 1.14, W 3/9 = 1.2, A = 2.29.

Thorax

1.50 mm long, 0.65 mm broad, blackish brown, undifferentiated, laterally compressed, lyre shaped; scutum black, without scales; scutellum (Figure 29) 0.45 mm long, 0.30 mm broad; rounded, brownish; sternopleuron and mesepimeron triangular, brownish.

Forewing (Figure 30)

3.45 mm long, 0.70 mm broad, wing with pale marking; yellowish scales present on veins; subcosta straight, 2.85

mm long, reaching costa; Base of costa with an pale interruption just external to cross vein; media straight, 2.70 mm long; radius straight, slightly curved at apex; cubitus bifurcated; anal vein 1.69 mm long; Vein 3 (R4+5) mainly dark; one or two fringe spots presents at vein 4–2 and 5–1.

Halter (Figure 31)

0.20 mm long, 0.09 mm broad, brownish tubular shaped, without scales or hairs, expanded at tip.

Hind Leg (Figure 32)

6.55 mm long, yellowish, longer than body; coxa 0.25 mm long, 0.20 mm broad, yellowish; trochanter 0.19 mm long, 0.13 mm broad, rounded, yellowish; femur 1.80 mm long, cylindrical, yellowish, covered with scales; femora not speckled, hind femur without knee spot at distal end; tibia 1.90 mm long, yellowish, not speckled, tibia bristles present in between joints; tarsus 2.41 mm long, yellowish, scally, five segmented, 1st tarsal segment 1.05 mm long, 2nd tarsal segment 0.72 mm long, 3rd tarsal segment 0.33 mm long, 4th tarsal segment 0.20 mm long, 5th tarsal segment 0.10 mm long, tarsomeres without bands.

Other Legs

Special marks: similar.

Abdomen (Figure 33)

1.75 mm long, 0.70 mm broad, reddish brown, without banded, dorsal plate reddish; post genital plate 0.10 mm long, 0.09 mm broad, brownish, hairy, longer than anal cerci; post genital plate and anal cerci in right angle; anal cerci 0.05 mm long, 0.04 mm broad, densely hairy, brown.

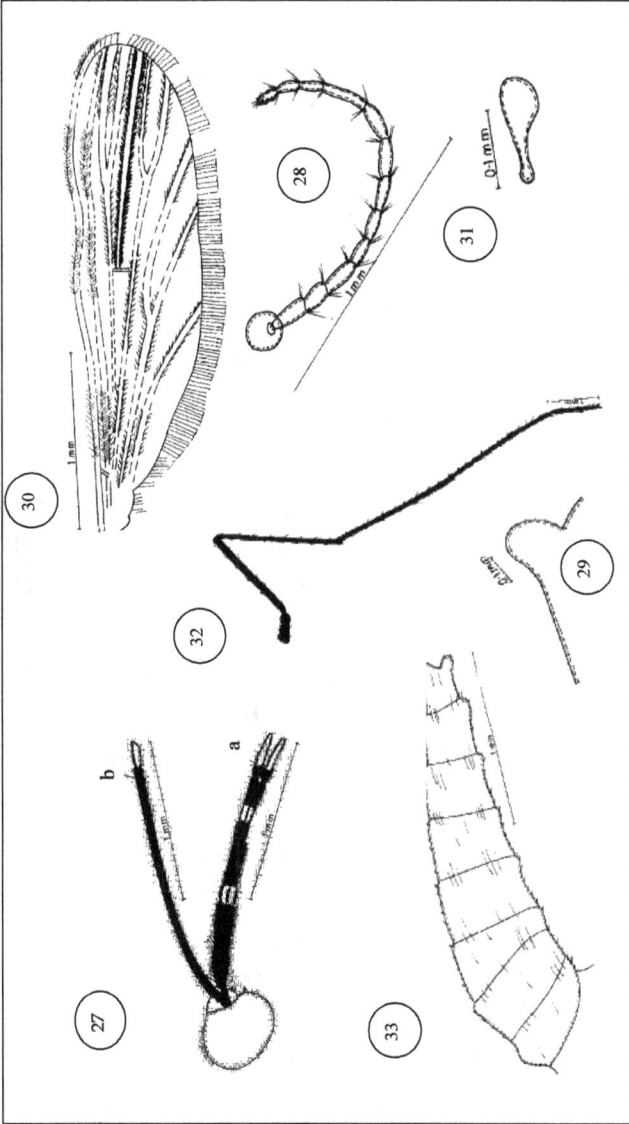

Plate 5–*Anopheles* (Cellia) *culicifacies* sp. nov.

Figure 27: Head -Palpi (1a) and Proboscis (1b); **Figure 28:** Antenna; **Figure 29:** Scutellum; **Figure 30:** Forewing; **Figure 31:** Halter; **Figure 32:** Hind leg; **Figure 33:** Abdomen

Colour

Black	:	Eyes.
Blackish brown	:	Head, thorax.
Yellowish	:	Proboscis labium wing, legs.
Brownish	:	Antenna, halter.
Reddish brown	:	Abdomen.
Male	:	3.25 mm long, slender, smaller than female; antenna Plumose; phytophagus.
Host	:	Human and cattle.
Host plant	:	Unknown.
Holotype	:	Female, India, Maharashtra, Patan coll. Jagtap, M.B., 11-VII-2009 head, antenna, hind leg, wing and abdomen mounted on the slide, labeled as above.
Paratype	:	135 ♂, 376 ♀, sex ratio (M:F) 1:2.78 coll. Jagtap, M.B. Jan. 2006 to Dec. 2009.

Distributional Record

2 ♂ 8♀ Jath, 5-III-2006; 3 ♂ 9 ♀ Saswad, 12-III-2006; 4 ♂ 7 ♀ Mahabaleshwar, 23-IV-2006; 3 ♂ 8 ♀ Kolhapur, 24-VI-2006; 4♂ 10 ♀ Jaysingpur, 9-VII-2006; 6 ♂ 17 ♀ Miraj, 11-VI-2006; 6 ♂ 11 ♀ Kolhapur, 24-VI-2006; 6 ♂ 10 ♀ Kagal, 12-VIII-2006; 5 ♂ 13 ♀ Malakapur, 11-II-2007; 4 ♂ 14♀ Koregaon, 3-V-2007; 4 ♂ 9 ♀ Shirala, 14-VI-2007; 9 ♂ 26 ♀ Pune, 25-VII-2007; 4 ♂ 9 ♀ Junner, 16-VIII-2007; 4 ♂ 10 ♀ Satara, 23-XI-2007; 7 ♂, 18 ♀, Wai, 9-II-2008; 3 ♂ 7 ♀ Vita,

23-III-2008; 5 ♂ 8 ♀ Ajara, 26-IV-2008; 5 ♂ 21 ♀ Kagal, 12-VII-2008; 5 ♂ 9 ♀ Medha, 20-VII-2008; 4 ♂ 13 ♀ Bhor, 10-VIII-2008; 4 ♂ 10 ♀ Mhaswad, 23-VIII-2008; 5 ♂ 9 ♀ Saswad, 11-X-2008; 3 ♂ 15 ♀ Junner, 22-II-2009; 4 ♂ 18 ♀ Vita, 25-IV-2009; 4 ♂ 18♀ Kolhapur, 14-VI-2009; 4 ♂ 19 ♀ Patan, 11-VII-2009; 4 ♂ 1 6 ♀ Wai, 9-VIII-200; 9 ♂ 18 ♀ Bhor, 29-XI-2009; 5 ♂ 16♀ Jaysingpur, 12-XII-2009.

Remark

According to the key of Rao (1984) this species runs close to *Anopheles culicifacies* by following characters.

1. Apical pale band nearly equal to the pre-apical dark band.

2. Tarsomeres without bands.

3. Vein 3 (R4+5) mainly dark.

4. Innercosta interrupted.

However, some following additional characters have been observed,

1. Scutellum size (0.45 mm long and 0.30 mm broad) and shape (rounded).

2. Flagellar formula: 3 L/W = 3.33, 9 L/W = 3.5, L 3/9 = 1.14, W 3/9 = 1.2, A = 2.29.

3. Phyllogenetically it runs close to *Anopheles culicifacies*. However, it differs from 16 species by having neighborhood joining branch length = 0.27758278.

Anopheles (Cellia) *mahabaleshwari* sp. nov

Female (Figure 56)

4.93 mm long, 0.80 mm broad; head 0.70 mm long, 0.53 mm broad, blackish brown; antenna 1.08 mm long,

Mosquito Diversity and Control

brownish; thorax 1.60 mm long, 0.80 broad, blackish brown; forewing 3.40 mm long, 0.70 mm broad, yellowish; hind leg 8.40 mm long, yellowish; abdomen 2.80 mm long, 0.70 mm broad, reddish brown.

Head (Figure 35)

0.53 mm long, 0.60 mm broad, blackish brown, globular; white narrow rod like scales; compound eyes black, rounded, ocular space 0.35 mm, interocular distance 0.25 mm long; vertex smooth, dark brown; proboscis 2.60 mm long, blackish, scally, cylindrical; labium 0.40 mm long, brown, yellowish, scally; labellum 2.03 mm long, brownish, scally; palpi 2.20 mm long, 5 segmented, slender, as long as proboscis, densely scally, apical pale band narrow and preapical pale band is long 1.20 mm; mandibles and maxillae well developed.

Antenna (Figure 36)

1.06 mm long, 15 segmented, hairy, brownish, pilose; scape 0.05 mm long, 0.15 mm broad, brownish; pedicel 0.26 mm long, 0.10 mm broad, yellowish brown; flagellum 0.80 mm long, 13 segmented and 3 pale bands.

Flagellar Formula

1 L/W = 4, 13 L/W = 2.4, L1/13 = 1, W1/13 = 0.6, A = 4.

Thorax

1.60 mm long, 0.80 mm broad, blackish brown, undifferentiated, laterally compressed, lyre shaped; scutum black, hair like scales are present; scutellum (Figure 37) 0.4 mm long, 0.3 mm broad rounded, globular and opaque, brownish; sternopleuron and mesepimeron triangular, brownish.

Forewing (Figure 38)

3.40 mm long, 0.70 mm broad, wing with pale marking, wide dark pale band, golden, scales present on veins; spotting of wings on costa and vein 1 only; subcosta straight, 2.85 mm long, reaching costa, only one dark spots on costa; radius straight, slightly curved at apex; cubitus bifurcated.

Halter (Figure 39)

0.50 mm long, 0.09 mm broad, yellowish, tubular, dumb-bell shaped, without scales or hairs, expanded at tip.

Hind Leg (Figure 40)

8.40 mm long, yellowish, longer than body; coxa 0.25 mm long, 0.10 mm broad, yellowish; trochanter 0.15 mm long, 0.10 mm broad, rounded, yellowish; femur 2.10 mm long, cylindrical, yellowish, covered with scales; femora not speckled, hind femur without knee spot at distal end; tibia 2.30 mm long, yellowish, speckling is observed in all leg, hind femur with not distinct knee spot; tibia bristles present in between joints; tarsus 2.60 mm long, yellowish, scally, five segmented, 1st tarsal segment 1.10 mm long, 2nd tarsal segment 0.85 mm long, 3rd tarsal segment 0.35 mm long, 4th tarsal segment 0.20 mm long, 5th tarsal segment 0.10 mm long, tip of tarsi not white, tarsi of front leg without pale lines. The proportion of femur and 1st tarsal segment is 2:1.

Other Legs

Special marks: similar.

Abdomen (Figure 41)

2.80 mm long, 0.70 mm broad, reddish brown, without banded, dorsal plate reddish; post genital plate 0.90 mm long, 0.07mm broad, brownish, hairy, longer than anal cerci, post genital plate and anal cerci in right angle; anal cerci 0.01 mm long, 0.03 mm broad, densely hairy, brown.

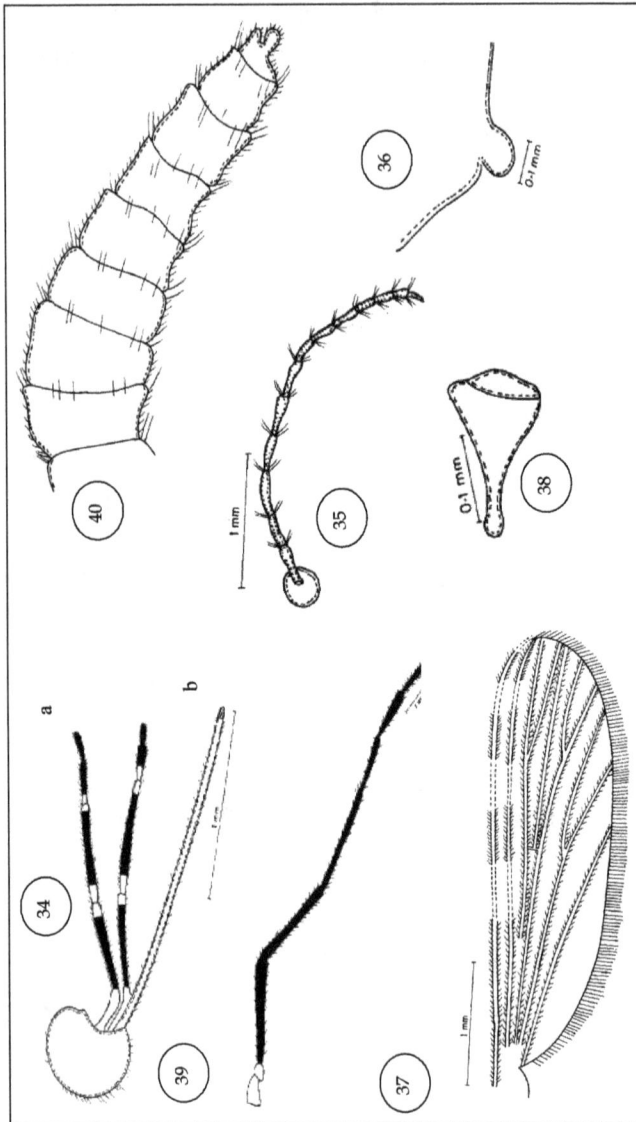

Plate 6–*Anopheles* (Cellia) *mahabaleshwari.* sp. nov.
Figure 34: Head-Palpi (1a) and Proboscis (1b); **Figure 35**: Antenna; **Figure 36**: Scutellum; **Figure 37**: Forewing; **Figure 38**: Halter; **Figure 39**: Hind leg; **Figure 40**: Abdomen

Colour

Black	: Eyes.
Blackish brown	: Head, thorax.
Yellowish	: Proboscis, labium wing, legs.
Brownish	: Antenna, halter.
Reddish brown	: Abdomen.
Male	: 3.35 mm long, slender, smaller than female; antenna plumose, phytophagus.
Host	: Human and cattle.
Host plant	: Unknown.
Holotype	: Female, India, Maharashtra, Mahabaleshwar coll. Jagtap, M.B., 10-II-2008 head, antenna, hind leg, wing and abdomen mounted and pinned, labelled as above.
Paratype	: 17 ♂, 49 ♀, sex ratio (M:F) 1:2.88 coll. Jagtap, M.B. Aug. 2006 to Nov. 2009.
Etymology	: The species name *Anopheles mahabaleshwari* sp. nov. refers to the collection site of mosquitoes, *i.e.* Mahabaleshwar, district Satara, Maharashtra, India.

Distributional Record

1 ♂ 5 ♀ Kagal, 12-VIII-2006; 0 ♂ 2 ♀ Miraj, 14-III-2007; 2♂ 6 ♀ Patan, 21-VI-2007; 4 ♂ 8 ♀ Shirala, 13-IX-2007; 2 ♂ 5 ♀ Satara, 23-XI-2007; 2 ♂ 5 ♀ Pune, 12-I-2008; 5 ♂ 13 ♀ Mahabaleshwar, 10-II-2008; 1 ♂ 4 ♀ Patan, 11-VII-2009; 0 ♂ 1 ♀ Baramati, 28-XI-2009.

Remark

According to the key of Rao (1984) this species runs close to *Anopheles* (Cellia) *dthali* Patton 1905 by following characters.

1. Spotting of wings confined to costa and vein 1 only.
2. Head scales narrow and rod like.

However, it differs from the above species by having following characters:

1. Scutellum globular and opaque, 0.4 mm long and 0.3 mm broad.
2. Wing area dark except costa and vein 1.
3. Preapical pale band is long.
4. The proportion of femur and 1st tarsal segment is 2:1.
5. Flagellar formula :
   1 L/W = 4, 13 L/W = 2.4, L1/13 = 1, W1/13 = 0.6, A = 4.

Anopheles (Cellia) *waii* sp. nov.

Female (Figure 57)

4.61 mm long, 1.24 mm broad; head 0.53 mm long, 0.60 mm broad, blackish brown; antenna 1.70 mm long, brownish; thorax 1.12 mm long, 1.24 broad, blackish brown; forewing 3.09 mm long, 0.85 mm broad, yellowish; hind leg 8.00 mm long, yellowish with speckling; abdomen 2.96 mm long, 0.65 mm broad, reddish brown.

Head (Figure 41)

0.53 mm long, 0.60 mm broad, blackish brown, globular, white narrow scales; compound eyes black, rounded, ocular space 0.35 mm, interocular distance 0.25 mm long; vertex smooth, white scales are present on scales; proboscis 1.72 mm long, apical pale half of proboscis is pale, preapical pale band is short, scally, cylindrical; labium 0.35 brown, yellowish, scally; labellum 1.37 mm long, brownish, scally; palpi 1.70 mm long, with 4 pale bands, one narrow and 3 broad pale bands, 5 segmented, slender, as long as proboscis, densely scally, apical pale band narrow and preapical pale bands is short; mandibles and maxillae well developed.

Antenna (Figure 42)

1.70 mm long, 15 segmented, hairy, brownish, pilose; scape 0.05 mm long, 0.15 mm broad, brownish; pedicel 0.25 mm long, 0.10 mm broad, yellowish brown; flagellum 1.45 mm long, 13 segmented and 3 pale bands.

Flagellar Formula

1 L/W = 4, 13 L/W = 2.4, L1/13 = 1, W1/13 = 0.6, A = 4.

Thorax

1.12 mm long, 1.24 mm broad, blackish brown, undifferentiated, laterally compressed, lyre shaped, scutum black, with white scales, white spots on thorax; scutellum (Figure 43) 0.5 mm long, 0.6 mm broad; rounded, globular and opaque, brownish; sternopleuron and mesepimeron triangular, brownish.

Forewing (Figure 44)

3.09 mm long, 0.85mm broad, wing with 4 pale marking and wide dark pale band, golden, scales present on veins; subcosta straight, 2.75 mm long, reaching costa, only one dark spots on costa; radius straight, slightly curved at apex; cubitus bifurcated.

Halter (Figure 45)

0.45 mm long, 0.08 mm broad, yellowish tubular dumb-bell shaped, without scales or hairs, expanded at tip.

Hind Leg (Figure 46)

8.00 mm long, yellowish, longer than body; coxa 0.25 mm long 0.15 mm broad, yellowish; trochanter 0.15 mm long, 0.10 mm broad, rounded, yellowish; femur 1.95 mm long, cylindrical, yellowish, covered with scales; femora and tibia speckled, tibia 2.45 mm long, yellowish, speckling is observed in all leg, hind leg without band at tibia tarsal joints, tibial bristles present in between joints; tarsus 3.20 mm long, yellowish, scally, five segmented, 1st tarsal segment 1.25 mm long, 2nd tarsal segment 1.05 mm long, 3rd tarsal segment 0.60 mm long, 4th tarsal segment 0.20 mm long, 5th tarsal segment 0.10 mm long and completely dark.

Other Legs

Special marks: similar.

Abodmen (Figure 47)

2.96 mm long, 0.65 mm broad, reddish brown, without banded, dorsal plate reddish; post genital plate 0.90 mm long, 0.07mm broad, brownish, hairy, longer than anal cerci, post genital plate and anal cerci in right angle; anal cerci 0.02 mm long, 0.03 mm broad, densely hairy, brown.

Colour

Black	: Eyes.
Blackish brown	: Head, thorax.
Yellowish	: Proboscis labium wing, legs.
Brownish	: Antenna, halter.

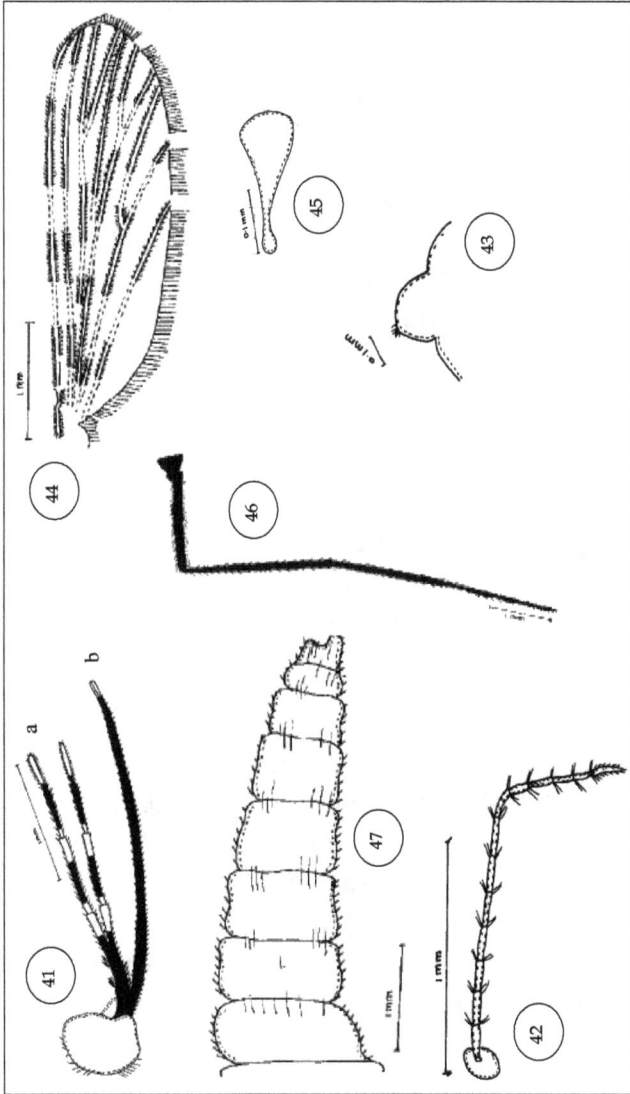

Plate 7–*Anopheles* (Cellia) *waii* sp. nov.

Figure 41: Head-Palpi (1a) and Proboscis (1b); **Figure 42:** Antenna; **Figure 43:** Scutellum; **Figure 44:** Forewing; **Figure 45:** Halter; **Figure 46:** Hind leg; **Figure 47:** Abdomen

Reddish brown	: Abdomen.
Male	: 3.45 mm long, slender, smaller than female; antenna plumose, phytophagus.
Host	: Human and cattle.
Host plant	: Unknown.
Holotype	: Female, India, Maharashtra. Wai coll. Jagtap, M.B., 09-II-2008 head, antenna, hind leg, wing and abdomen mounted and pinned, labeled as above.
Paratype	: 34 ♂, 82 ♀, sex ratio (M:F) 1:2.41coll. Jagtap, M.B. Mar. 2006 to Nov. 2009.
Etymology	: The species name *Anopheles waii* sp. nov. refers to the collection site of mosquitoes *i.e.* Wai, district Satara, Maharashtra, India.

Distributional Record

3 ♂ 6 ♀ Jath, 5-III-2006; 2♂ 6 ♀ Miraj, 11-VI-2006; 3 ♂ 7 ♀ Malkapur, 1-II-2007; 1♂ 5 ♀ Shirala, 14-VI-2007; 2 ♂ 5 ♀ Miraj, 21-IX-2007; 2 ♂ 3 ♀ Pune, 20-XII-2007; 3 ♂, 5 ♀ Baramati, 13-I-2008; 4 ♂ 11 ♀ Wai, 09-II-2008; 4 ♂ 9 ♀ Kagal, 12-VII-2008; 3 ♂ 6 ♀ Saswad, 11-X-2008; 1 ♂ 3 ♀ Bhor, 21-II-2009; 4 ♂ 7 ♀ Patan, 11-VII-2009; 1♂ 7 ♀ Junner, 22-VIII-2009; 1 ♂ 2 ♀ Bhor, 29-XI-2009.

Discussion

According to the key of Rao (1984) this species runs close to *Anopheles* (Cellia) *tasselattus* Theobald 1901 by following characters.

1. Legs speckled.
2. Palpi with 4 pale bands, one narrow and 3 broad pale bands.
3. Apical half of proboscis is pale.
4. Hind leg tarsomere completely dark.

However, it differs from the above species by having following characters

1. Scutellum globular and opaque, 0.4 mm long and 0.3 mm broad.
2. Preapical pale bands short.
3. Flagellar formula:
   1 L/W = 4, 13 L/W = 2.4, L1/13 = 1, W1/13 = 0.6, A = 4.

Anopheles (Cellia) *karveeri* sp. nov.

Female (Figure 58)

3.80 mm long, 1.20 mm broad; head 0.60 mm long, 0.50 mm broad, blackish brown; antenna 1.90 mm long, brownish; thorax 1.30 mm long, 1.20 mm broad, blackish brown; forewing 3.40 mm long, 0.75 mm broad, yellowish; hind leg 7.40 mm long, yellowish; abdomen 1.90 mm long, 0.75 mm broad, reddish brown.

Head (Figure 48)

0.60 mm long, 0.50 mm broad, blackish brown, globular, white narrow scales; compound eyes black, rounded, ocular space 0.20 mm, interocular distance 0.15 mm long; vertex smooth, dark brown; proboscis 1.30 mm long, blackish, scally, cylindrical; labium 0.25 mm brown, yellowish, scally; labellum 1.05 mm long, brownish, scally; palpi 1.20 mm long with pale tip and not equal to proboscis, 5 segmented, slender, as long as proboscis, densely scally, apical pale band narrow and preapical pale bands is long; mandibles and maxillae well developed.

Antenna (Figure 49)

1.90 mm long, 15 segmented, hairy, brownish, pilose; scape 0.05 mm long, 0.15 mm broad, brownish; pedicel 0.25 mm long, 0.10 mm broad, yellowish brown; flagellum 1.65 mm long, 13 segmented with 3 pale bands.

Flagellar Formula

3 L/W = 2, 9 L/W = 2.5, L 3/9 = 1.2, W 3/9 = 1.5, A = 1.8.

Thorax

1.30 mm long, 1.2 mm broad, blackish brown, undifferentiated, laterally compressed, lyre shaped; scutum black, without scales, white spots on thorax; scutellum (Figure 50) 0.5 mm long, 0.6 mm broad, rounded, globular and opaque, brownish; sternopleuron and mesepimeron triangular, brownish.

Forewing (Figure 51)

3.40 mm long, 0.75 mm broad, wing with 4 dark band, golden, scales present on veins; vein 3 mainly dark, subcosta straight, 2.60 mm long, reaching costa, pale spots present on other wing veins besides costa and vein 1; fring spot on vein 4.1, 5.2 and 6, radius straight, slightly curved at apex; cubitus bifurcated.

Halter (Figure 52)

0.3 mm long, 0.12 mm broad, yellowish tubular dumb-bell shaped, without scales or hairs, expanded at tip.

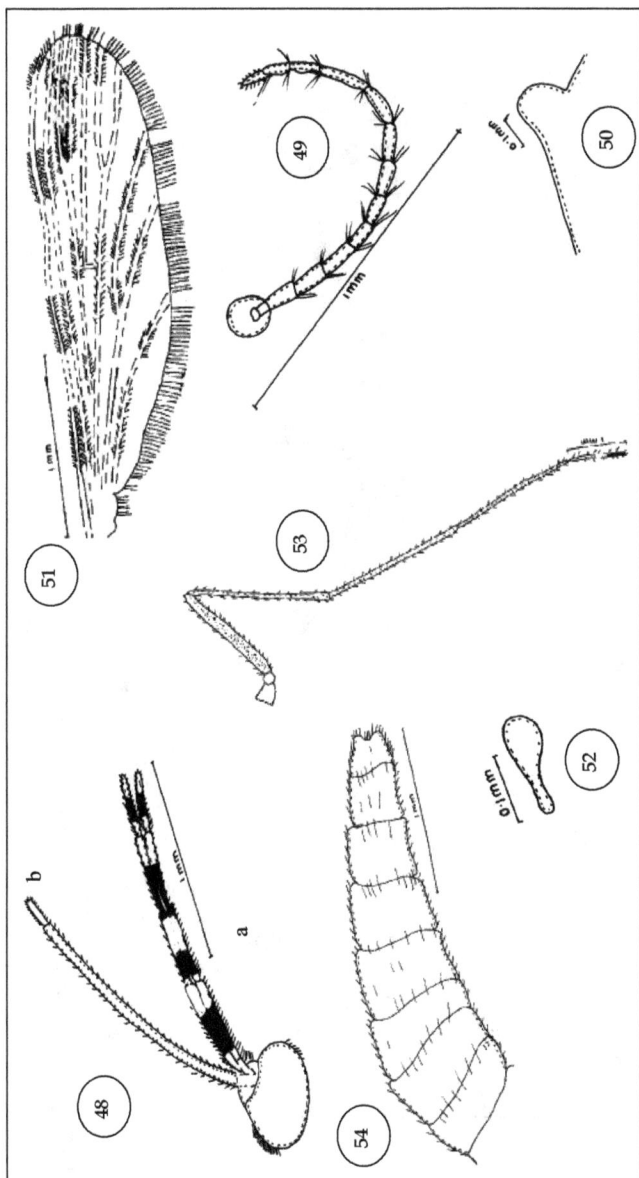

Plate 8–*Anopheles* (Cellia) *karveeri* sp. nov.

Figure 48: Head-Palpi (a) and Proboscis (b); **Figure 49:** Antenna; **Figure 50:** Scutellum; **Figure 51:** Forewing; **Figure 52:** Halter; **Figure 53:** Hind leg; **Figure 54:** Abdomen

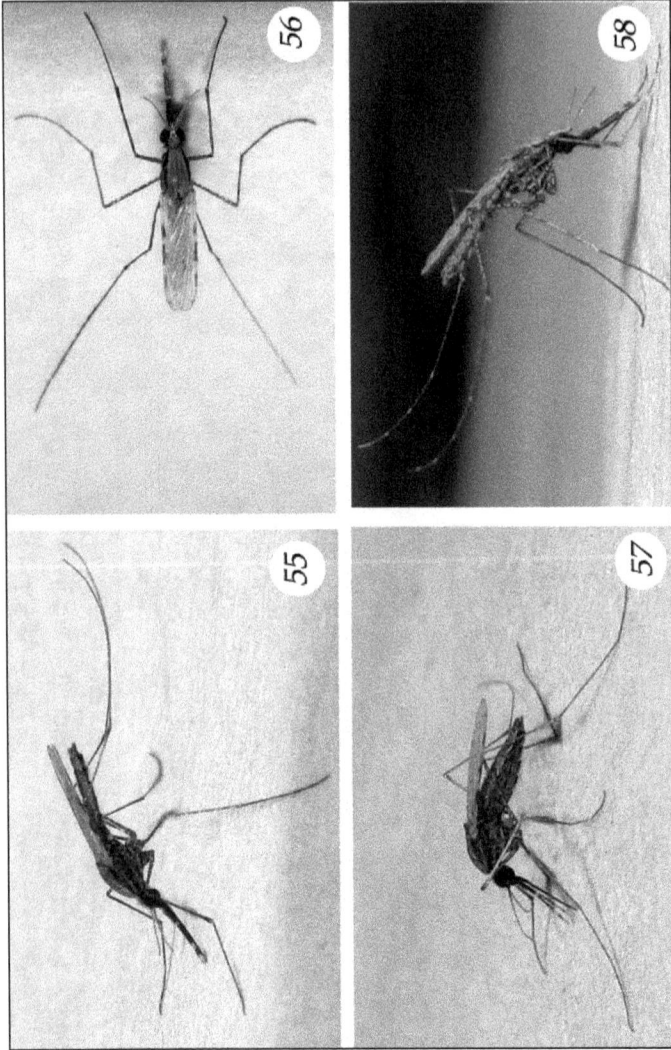

Plate 9–Figure 55: *Anopheles* (Cellia) *culicifacies* sp. nov. **Figure 56**: *Anopheles* (Cellia) *mahabaleshwari.* sp. nov.; **Figure 57**: *Anopheles* (Cellia) *waii* sp. nov.; **Figure 58**: *Anopheles* (Cellia) *karveeri* sp. nov.

Hind Leg (Figure 53)

7.40 mm long, yellowish, longer than body; coxa 0.45 mm long 0.20 mm broad, yellowish; trochanter 0.25 mm long, 0.15 mm broad, rounded, yellowish; femur 1.40 mm long, cylindrical, yellowish, covered with scales; femora and tibia not speckled, hind femur without knee spot at distal end; tibia 2.40 mm long, yellowish, hind femur with not distinct knee spot, tibia bristles present in between joints; tarsus 2.90 mm long, yellowish, scally, fore tarsi unbanded, five segmented, 1st tarsal segment 1.05 mm long, 2nd tarsal segment 0.85 mm long, 3rd tarsal segment 0.55 mm long, 4th tarsal segment 0.35 mm long, 5th tarsal segment 0.10 mm long, hind tarsomere not white.

Other Legs

Special marks: similar.

Abdomen (Figure 54)

1.90 mm long, 0.75 mm broad, reddish brown, without banded; dorsal plate reddish, brownish, hairy, longer than anal cerci, post genital plate and anal cerci in right angle; anal cerci 0.05 mm long, 0.05 mm broad, densely hairy, brown.

Colour

Black	: Eyes.
Blackish brown	: Head, thorax.
Yellowish	: Proboscis labium wing, legs.
Brownish	: Antenna, halter.
Reddish brown	: Abdomen.
Male	: 3.10 mm long, slender, smaller than female; antenna plumose, phytophagus.

Host	: Human and cattle.
Host plant	: Unknown.
Holotype	: Female, India, Maharashtra. Kolhapur coll. Jagtap, M.B., 24-VI-2006 head, antenna, hind leg, wing and abdomen mounted and pinned, labeled as above.
Paratype	: 84 ♂, 237 ♀, sex ratio (M:F) 1:2.82 coll. Jagtap, M.B. Jun. 2006 to Dec. 2009.
Etymology	: The species name *Anopheles karveeri* sp. nov. refers to the collection site of mosquitoes *i.e.* Karveer tahsil, district Kolhapur, Maharashtra, India.

Distributional Record

9♂ 18 ♀ Miraj, 11-VI-2006; 4 ♂ 15 ♀ Kolhapur, 24-VI-2006; 7 ♂ 18 ♀ Kagal, 12-VIII-2006; 2♂ 5 ♀ Koregaon, 3-V-2007; 5♂ 13 ♀ Shirala, 14-VI-2007; 8♂ 20♀, Pune, 25-VII-2007; 5♂ 11 ♀ Junner, 16-VIII-2007; 5♂ 13 ♀ Satara, 23-XI-2007; 2♂ 5♀ Wai, 9-II-2008; 0 ♂ 3♀ Vita, 23-III-2008; 1 ♂ 4 ♀ Ajara, 26-IV-2008; 5 ♂ 18 ♀ Kagal, 12-VII-2008; 8 ♂ 15 ♀ Medha, 20-VII-2008; 4 ♂ 10 ♀ Mhaswad, 23-VIII-2008; 2♂ 8 ♀ Saswad, 11-X-2008; 2♂ 11 ♀ Junner, 22-II-2009; 3♂ 10 ♀ Vita, 25-IV-2009; 2♂ 7♀ Kolhapur, 14-VI-2009; 4 ♂ 12 ♀ Patan, 11-VII-2009; 2 ♂ 10 ♀ Bhor, 29-XI-2009; 4 ♂ 13♀ Jaysingpur, 12-XII-2009.

Remark

According to the key of Rao (1984) this species runs close to *Anopheles* (Cellia) *culicifascies* Giles 1901 by following characters.

1. Femur and tibia not speckled.
2. Pale spots present on other wing veins besides costa and vein 1.
3. Palpi with pale tip.

However, it differs from the above species by having following characters

1. Palpi not equal to proboscis.
2. Fringe spot on vein 4.1, 5.2 and 6.
3. Scutellum globular and opaque, 0.5 mm long and 0.6 mm broad.
4. Flagellar formula:
   3 L/W = 2, 9 L/W = 2.5, L 3/9 = 1.2, W 3/9 = 1.5, A = 1.8.

Anopheles (Cellia) *krishnai* sp.nov.

Female (Figure 80)

4.93 mm long, 0.80 mm broad; head 0.42 mm long, 0.53 mm broad, blackish brown; antenna 1.08 mm long, brownish; thorax 1.60 mm long, 0.80 broad, blackish brown; forewing 3.40 mm long, 0.70 mm broad, yellowish; hind leg 8.40 mm long, yellowish; abdomen 2.80 mm long, 0.70 mm broad, reddish brown.

Head (Figure 59)

0.42 mm long, 0.48 mm broad, blackish brown, globular, white narrow scales; compound eyes black, rounded, ocular space 0.25 mm, interocular distance 0.15 mm long; vertex smooth, dark brown; proboscis 1.93 mm long, blackish, scally, cylindrical; labium 0.20 mm brown, yellowish, scally; labellum 1.73 mm long, brownish, scally; palpi 1.90 mm long (Figure 1a), 5 segmented, slender, as long as proboscis, densely scally, 3 pale bands, apical pale band narrow and preapical pale bands is long, mandibles and maxillae well developed; phyrengeal armigers absent.

Antenna (Figure 60)

1.84 mm long, 15 segmented, hairy, brownish, pilose; scape 0.05 mm long, 0.15 mm broad, brownish; pedicel 0.20 mm long, 0.10 mm broad, yellowish brown; flagellum 1.64 mm long, 13 segmented and 3 pale bands.

Flagellar Formula

3 L/W = 3, 9 L/W = 2, L 3/9 = 1.5, W 3/9 = 1, A = 1.87.

Thorax

1.12 mm long, 0.84 mm broad, blackish brown, undifferentiated, laterally compressed, lyre shaped, scutum black, without scales, white spots on thorax; scutellum (Figure 61) 0.46 mm long, 0.6 mm broad; rounded, globular and opaque, brownish; sternopleuron and mesepimeron triangular, brownish.

Forewing (Figure 62)

3.09 mm long, 0.85 mm broad, wing with pale marking wide dark pale band, golden, scales present on veins; subcosta straight, 2.30 mm long, reaching costa, only one dark spots on costa; radius straight, slightly curved at apex; cubitus bifurcated. Wing with 4 dark marking on costa and vein 1. 6th vein with 3 dark spots.

Halter (Figure 63)

0.40 mm long, 0.07 mm broad, yellowish globular, broadly dumb-bell shaped, expanded at tip, without scales or hairs.

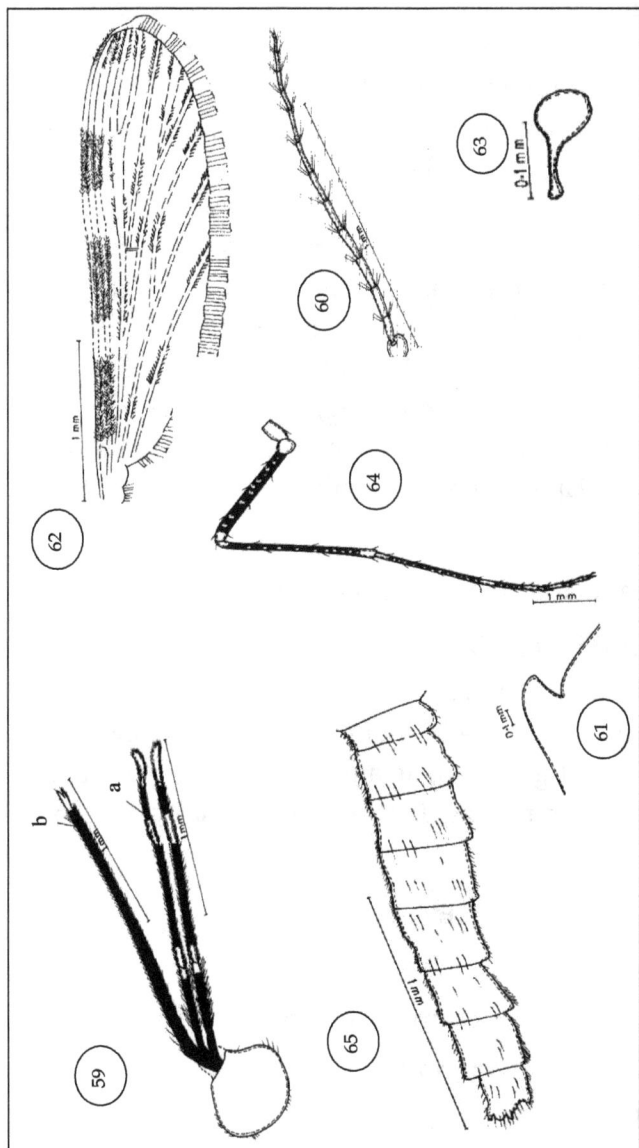

Plate 10—*Anopheles* (Cellia) *Krishnai* sp. nov.

Figure 59: Head-Palpi (1a) and Proboscis (1b); **Figure 60**: Antenna; **Figure 61**: Scutellum; **Figure 62**: Forewing; **Figure 63**: Halter; **Figure 64**: Hind leg; **Figure 65**: Abdomen

Hind Leg (Figure 64)

8.10 mm long, yellowish, longer than body; coxa 0.25 mm long 0.15 mm broad, yellowish; trochanter 0.15 mm long, 0.10 mm broad, rounded, yellowish; femur 1.90 mm long, cylindrical, yellowish, covered with scales; femora not speckled, hind femur without knee spot at distal end; tibia 2.50 mm long, yellowish, speckling observed in all legs, hind femur with distinct knee spot; tibia bristles present in between joints; white spots on hind femur and tibia; tarsus 3.30 mm long, yellowish, scally, five segmented, 1st tarsal segment 1.30 mm long, 2nd tarsal segment 1.10 mm long, 3rd tarsal segment 0.60 mm long, 4th tarsal segment 0.20 mm long, 5th tarsal segment 0.10 mm long, pretarsus longer than others. Tip of hind tarsi white.

Other Legs

Special marks: more or less similar.

Abdomen (Figure 65)

2.24 mm long, 0.45 mm broad, reddish brown, without banded, dorsal plate reddish; brownish, hairy, longer than anal cerci, post genital plate and anal cerci in right angle; anal cerci 0.02 mm long, 0.03 mm broad, hairy, brown.

Colour

Black	:	Eyes.
Blackish brown	:	Head, thorax.
Yellowish	:	Proboscis, labium, wings, legs.
Brownish	:	Antenna, halter.
Reddish brown	:	Abdomen.
Male	:	3.25 mm long, slender, smaller than female; antenna plumose, phytophagus.

Host	: Human and cattle.
Host plant	: Unknown.
Holotype	: Female, India, Maharashtra, Wai coll. Jagtap, M.B., 09-II-2008 head, antenna, hind leg, wing and abdomen mounted and pinned, labeled as above.
Paratype	: 96 ♂, 254 ♀, sex ratio (M:F) 1:2.64 coll. Jagtap, M.B. Mar. 2006 to Nov. 2009.
Etymology	: The species name *Anopheles* (Cellia) *krishnai* sp. nov. refers to collection site of mosquitoes *i.e.* Wai city situated at the bank of river Krishna, hence the name.

Distributional Record

3 ♂ 13 ♀ Saswad, 12-III-2006; 4 ♂ 16 ♀ Kolhapur 24-VI-2006; 3♂, 11♀, Kagal, 12-VIII-2006; 1♂, 10♀, Jaysingpur, 12-XI-2006; 1 ♂, 8 ♀ Miraj 14-III-2007; 6♂ 13 ♀ Shirala, 14-VI-2007; 3♂ 10 ♀ Malakapur, 12-VII-2007; 13♂, 27♀, Kolhapur, 11-X-2007; 4♂ 8 ♀ Satara, 23-XI-2007; 6♂, 11♀, Wai, 9-II-2008; 1 ♂, 7♀, Vita, 23-III-2008; 2 ♂ 14 ♀ Ajara, 26-IV-2008; 4♂ 7 ♀ Saswad, 11-X-2008; 7 ♂ 13 ♀ Tasgaon, 8-XI-2008; 5♂ 11 ♀ Vita, 25-IV-2009; 3♂, 12♀, Satara, 27-VI-2009; 9 ♂ 19 ♀ Patan, 11-VII-2009; 4 ♂ 13 ♀ Ajara, 25-VII-2009; 3 ♂ 16 ♀ Wai, 9-VIII-2009; 11 ♂ 15 ♀ Junner, 22-VIII-2009; 3 ♂ 10 ♀ Kolhapur, 13-IX-2009.

Remark

According to Rao (1984) this species runs close to *Anopheles* (Cellia) *maculatus* by following characters.

1. Palpus long with 3 pale bands on palpi.
2. White spots on hind femur and tibia.

However, it differs from the above species by having following features

1. Scutellum globular and opaque, 0.46 mm long and 0.6 mm broad.
2. Speckling on legs
3. Preapical pale bands long.
4. Flagellar formula :
   $3 L/W = 3, 9 L/W = 2, L 3/9 = 1.5, W 3/9 = 1, A = 1.87$.
5. Phyllogenetically it runs close to *An. Funestus*. However, it differs from 19 species by having neighbourhood joining branch length = 0.295164.

Key to the Species of Subgenera of *Anopheles*

1. Wings entirely dark 2

   Wings with pale markings 4
2. Hind femur with distinct white knee
   spots at distal end .... *barianensis*

   Hind femur without such a knee spot 3
3. Head scales very narrow, rod
   like .... *Aitkeni, bengalensis,*
   .... *insulaeflorum pinjaurensis,*

   Head scales of ordinary type .... *Culiciformis sintoni*

   Tibia about twice the length of body .... *Atpadi* sp. nov.
4. Hind femur with an outstanding tuft
   of white and black scales as its distal
   end, visible to the naked eye ...... *annandalei, interruptus*

Hind femur not so 5

Bristles on hind coxa, femora about
twice the length of body .... *Ajarae* sp. nov.

5. Hind femur with a broad white .... *lindesayi* and,
.... Subspecies *nilgiricus*

Hind femur without such a band 6

6. Inner quarter of costa pale .... *gigas* and its
.... varieties
.... *simlensis, bayleyi*
.... *refutans* (Sri Lanka only)

White strip on thorax, dark pale
band on forewing, abdomen
narrow .... *kolhapuri* sp. nov.

Inner quarter of costa mainly dark
though there may be a few
scattered pale scales 7

7. Palpi with definite pale markings; .... *nigerrimus*
clypeus with tuft of black scales .... *sinensis*
at side. .... *Argyropus, crawfordi*
.... *nitidus, pediataenitus*

Palpi without any pale markings 8

8. Females with prominent tuft of .... *barbirostris*
scales on ventral side of .... *ahomi*
abdominal segment VII; inner .... *barumbrosus*
third of costa with a few pale .... *compestris*
scales.

Females without such a tuft .... *umbrosus*

Inner third of costa without .... *roperi*
pale scales.

Anopheles (Anopheles) *compestris* Reid 1962

Female (Figure 81)

4.55 mm long, 0.90 mm broad, comparatively large mosquito with shaggy appearance; head 0.85 mm long, 0.70 mm broad, blackish brown; antenna 1.90 mm long, brownish; thorax 1.80 mm long, 0.85 broad, blackish brown; forewing 3.70 mm long, 0.90 mm broad, yellowish; hind leg 7.57 mm long, yellowish; abdomen 1.90 mm long, 0.90 mm broad, reddish brown.

Head (Figure 66)

0.85 mm long, 0.70 mm broad, blackish brown, globular, with narrow scales; compound eyes black, rounded, ocular space 0.30 mm, interocular distance 0.21 mm long; vertex smooth, dark brown; nape 0.20 mm long, 0.18 mm broad, tubular, brownish; proboscis 2.60 mm long, entirely blackish, scally, cylindrical; labium 0.30 brown, yellowish, scally; labellum 2.30 mm long, brownish, scally; palpi 2.60 mm long and dark, 5 segmented, joint of palpi distinguishable, slender, as long as proboscis, densely scaly and dark, palpi without pale marking, mandibles and maxillae well developed, yellow.

Antenna (Figure 67)

1.90 mm long, 15 segmented, hairy, brownish, pilose; scape 0.06 mm long, 0.15 mm broad, brownish; pedicel 0.30 mm long, 0.13 mm broad, yellowish brown; flagellum 1.65 mm long, 13 segmented.

Flagellar Formula

3 L/W = 3.2, 9 L/W = 2.6, L 3/9 = 1.23, W 3/9 = 1, A = 2.00.

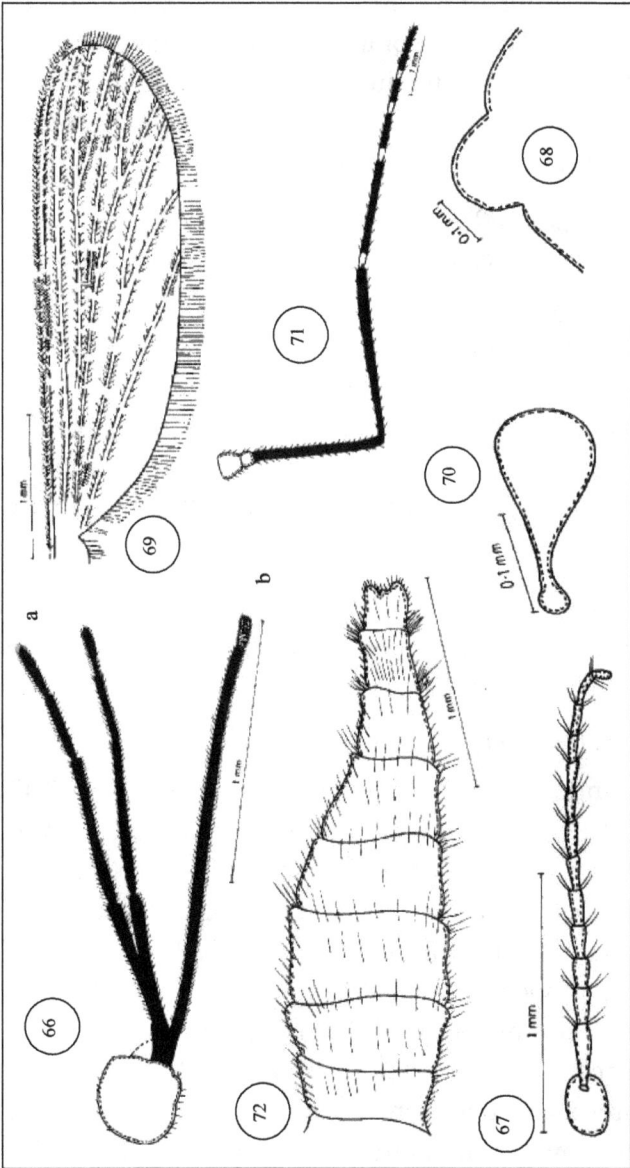

Plate 11–*Anopeles (Anopeles) compestris*

Figure 66: Head-Palpi (1a) and Proboscis (1b); **Figure 67**: Antenna; **Figure 68**: Scutellum; **Figure 69**: Forewing; **Figure 70**: Halter; **Figure 71**: Hind leg; **Figure 72**: Abdomen.

Thorax

1.80 mm long, 0.85 mm broad, blackish brown, undifferentiated, laterally compressed, lyre shaped; scutum black, without scales; scutellum (Figure 68) 0.50 mm long, 0.40 mm broad; rounded, brownish; sternopleuron and mesepimeron triangular, brownish.

Forewing (Figure 69)

3.70 mm long, 0.90 mm broad, wing with pale and dark spots on vein, yellowish, scales present on veins; subcosta straight, 2.95 mm long, reaching costa, few pale spots on inner costa; fringe spot on vein 3 (R4+5) and 5.2 (CCu-2), on vein 5 dark scales more than half; media straight, 2.80 mm long; radius straight, slightly curved at apex; cubitus bifurcated; anal vein 1.75 mm long, fringe spot on vein 2.1 absent.

Halter (Figure 70)

0.22 mm long, 0.15 mm broad, brownish tubular shaped, without scales or hairs, expanded at tip.

Hind Leg (Figure 71)

7.57 mm long, yellowish, longer than body; coxa 0.45 mm long 0.25 mm broad, yellowish; trochanter 0.22 mm long, 0.15 mm broad, rounded, yellowish; femur 2.05 mm long, cylindrical, yellowish, covered with scales; femora not speckled, no pale area on hind femur; tibia 2.10 mm long, yellowish, not speckled, hind femur without tuft of white and black scales at its distal ends, hind femur with not distinct knee spot and without broad white band, tibia bristles present in between joints; tarsus 2.75 mm long, yellowish brown, tarsomeres with broad white bands, scally, five segmented, 1st tarsal segment 1.10 mm long, 2nd tarsal segment 0.80 mm long, 3rd tarsal segment 0.35 mm long, 4th tarsal segment 0.30 mm long, 5th tarsal segment 0.20 mm long, pretarsus longer than other, pale bands are present on tarsal segments.

Other Legs

Special marks: similar.

Abdomen (Figure 72)

1.90 mm long, 0.90 mm broad, reddish brown, prominent tuft of dark scales on venter of the VII abdominal segment, at the middle part of venter white scales are more than *An. barbirostris* and few scales are present on lateral side, dorsal plate reddish; post genital plate 0.15 mm long, 0.10 mm broad, brownish, hairy, longer than anal cerci, post genital plate and anal cerci in right angle; anal cerci 0.05 mm long, 0.05 mm broad, densely hairy, brown.

Colour

Black	: Eyes, palpi, proboscis.
Blackish brown	: Head, thorax.
Yellowish	: Labium wing, legs.
Brownish	: Antenna, halter.
Reddish brown	: Abdomen.
Male	: 3.35 mm long, slender, smaller than female; antenna plumose, phyto-phagus.
Host	: Human and cattle.
Host plant	: Unknown.
Holotype	: Female, India, Maharashtra, Kolhapur coll. Jagtap, M.B., 8-VIII-

2007 head, antenna, hind leg, wing and abdomen mounted on the slide, labeled as above.

Paratype : 15 ♂, 32 ♀, sex ratio (M:F) 1:2.13 coll. Jagtap, M.B. Jan. 2006 to Nov. 2009.

Distributional Record

1 ♂ 5 ♀ Kolhapur 24-VI-2006, 2♂ 4 ♀ Jaysingpur, 9-VII-2006; 0 ♂ 1♀ Malkapur, 11- II-2007; 4 ♂, 7 ♀, Kolhapur, 8-VIII-2007; 3♂, 5 ♀, Kolhapur, 11-X-2007; 1 ♂ 2 ♀ Kagal, 12-VII-2008; 4 ♂ 8 ♀ Kolhapur, 13-IX-2009.

Remarks

According to the key of Rao (1984) this species is *Anopheles* (Anopheles) *compestris* Reid 1962 by following characters. This species is reported from India for the first time.

1. Palpi completely dark.

2. Proboscis is quite long, dark and hairy.

3. Inner costa with few pale scales.

4. Hind leg tarsomeres with white broad bands.

5. Inner quarter of costa mainly dark sometimes with few scattered scales.

6. Female with prominent tuft of dark scales on ventral side of abdominal segment VII.

7. Broad white scales on middle part of venter of abdomen are more than *An. barbirostris*.

8. On vein 5 dark scales more than half.

9. Scutellum size (0.50 mm long and 0.40 mm broad) and shape (rounded).

10. Halter 0.22 mm long, 0.15 mm broad with tubular shape.

11. Flagellar formula:
    3 L/W = 3.2, 9 L/W = 2.6, L 3/9 = 1.23, W 3/9 = 1, A = 2.00.

12. Phyllogenetically also it is confirmed as *Anopheles* (Anopheles) *compestris* with NHJ branch length = 7.19385402.

Anopheles (Anopheles) *kolhapuri* sp. nov.

Female (Figure 82)

3.90 mm long, 0.70 mm broad; head 0.70 mm long, 0.60 mm broad, blackish brown; antenna 1.80 mm long, brownish; thorax 1.50 mm long, 0.70 broad, blackish brown; forewing 3.40 mm long, 0.80 mm broad, marked spotted wings, yellowish; hind leg 6.70 mm long, yellowish; abdomen 1.70 mm long, 0.90 mm broad, reddish brown.

Head (Figure 73)

0.43 mm long, 0.56 mm broad, blackish brown, globular, with narrow scales; compound eyes black, rounded, ocular space 0.30 mm, interocular distance 0.20 mm long; vertex smooth, dark brown; proboscis 2.23 mm long, blackish, scally, cylindrical; labium 0.20 brown, yellowish, scally; labellum 2.03 mm long, brownish, scally; palpi 2.00 mm long, 5 segmented, slender, as long as proboscis, densely scally, palpi completely dark but small bands at joints.

Antenna (Figure 74)

1.1 mm long, 15 segmented, hairy, brownish, pilose; scape 0.05 mm long, 0.10 mm broad, brownish; pedicel 0.20

mm long, 0.11 mm broad, yellowish brown; flagellum 0.90 mm long, 13 segmented with 3 pale bands.

Flagellar Formula

1 L/W = 4, 13 L/W = 2.4, L1/13 = 1, W1/13 = 0.6, A = 4.

Thorax

1.70 mm long, 1.10 mm broad, blackish brown, undifferentiated, laterally compressed, lyre shaped; scutum black, without scales, white strip on thorax by having longitudinal black and lateral line; rest of thoracic scales are whitish; scutellum (Figure 75) 0.5 mm long, 0.46 mm broad; rounded, globular and opaque, brownish; sternopleuron and mesepimeron triangular, brownish.

Forewing (Figure 76)

3.10 mm long, 0.60 mm broad, wing with dark pale band 0.09 mm long and vein no.3 is dark, golden, scales present on veins; subcosta straight, 2.55 mm long, reaching costa, inner costa interrupted; media straight, 2.60 mm long; radius straight, slightly curved at apex; cubitus bifurcated; vein 6 with pale spot on outer half.

Halter (Figure 77)

0.40mm long, 0.09 mm broad, brownish tubular shaped, without scales or hairs, expanded at tip.

Hind Leg (Figure 78)

9.60 mm long, yellowish, longer than body; coxa 0.35 mm long 0.15 mm broad, yellowish; trochanter 0.15 mm long, 0.10 mm broad, rounded, yellowish; femur 2.10 mm long, cylindrical, yellowish, covered with scales; femora not speckled, mid femur with large pale spot; tibia 2.60 mm long, yellowish, not speckled; hind femur with not distinct

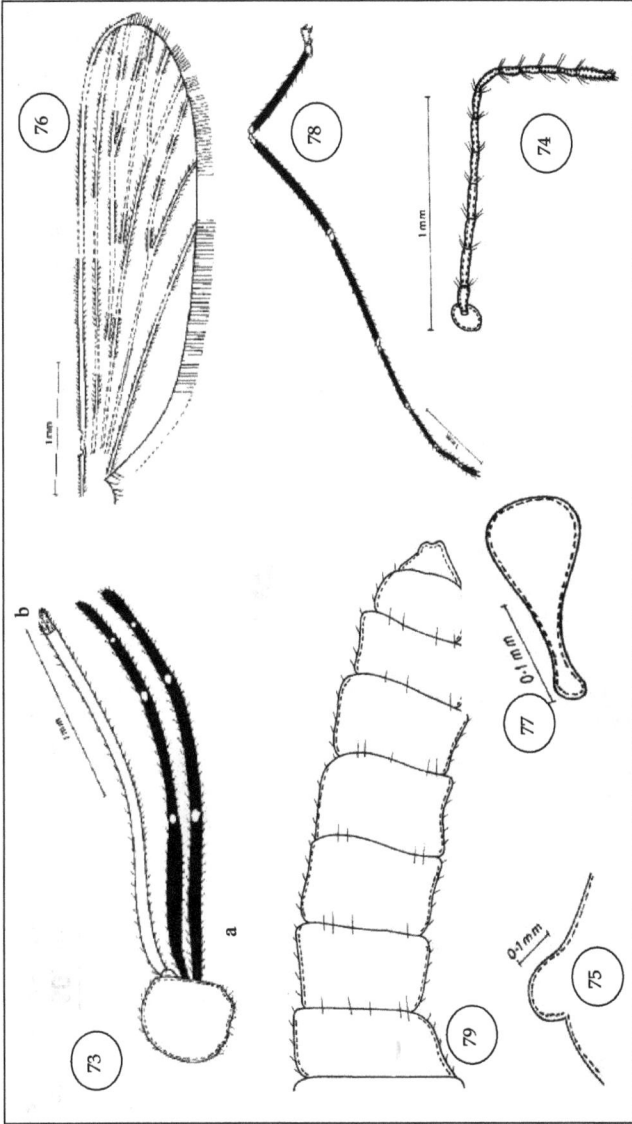

Plate 12–*Anopheles* (Anopheles) *kolhapuri* sp. nov.

Figure 73: Head-Palpi (a) and Proboscis (b); **Figure 74**: Antenna; **Figure 75**: Scutellum; **Figure 76**: Forewing; **Figure 77**: Halter; **Figure 78**: Hind leg; **Figure 79**: Abdomen.

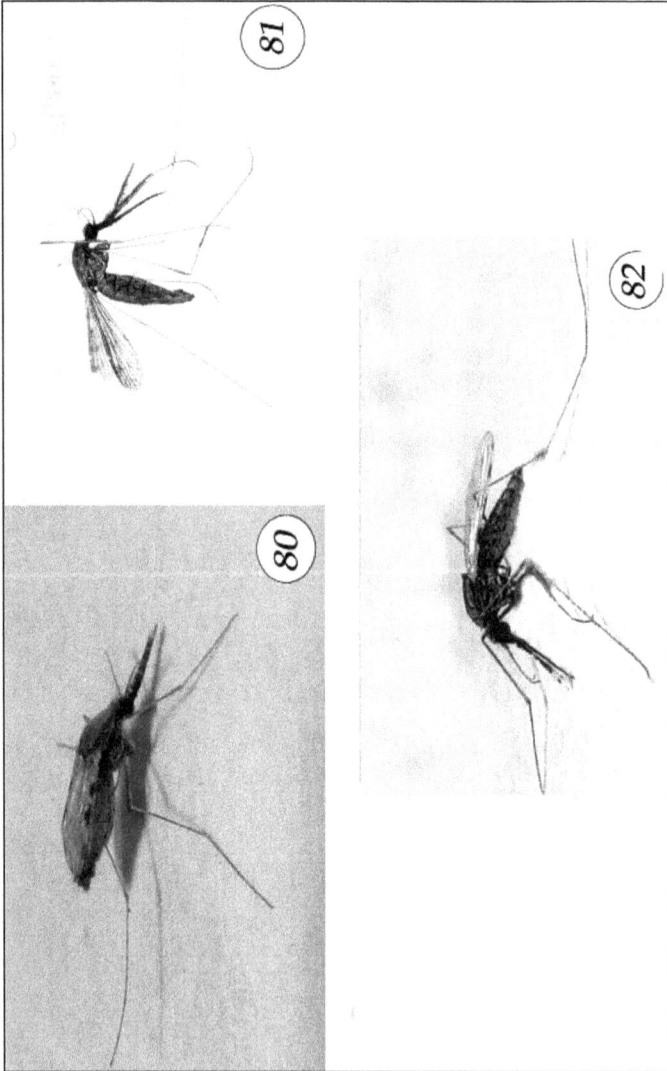

Plate 13—Figure 80: *Anopheles* (Cellia) *Krishnai* sp. nov.; **Figure 81**: *Anopheles* (Anopheles) *compestris*; **Figure 82**: *Anopheles* (Anopheles) *kolhapuri* sp. nov.

knee spot, tibia bristles present in between joints; tarsus 4.40 mm long, yellowish, scally, five segmented, 1st tarsal segment 1.95 mm long, 2nd tarsal segment 1.10 mm long, 3rd tarsal segment 0.80 mm long, 4th tarsal segment 0.35 mm long, 5th tarsal segment 0.20 mm long, pretarsus longer than other.

Other Legs

Special marks: similar.

Abdomen (Figure 79)

3.20 mm long, 0.90 mm broad, reddish brown, without banded, dorsal plate reddish; post genital plate 0.90 mm long, 0.08 mm broad, brownish, hairy, longer than anal cerci, post genital plate and anal cerci in right angle; anal cerci 0.05 mm long, 0.04 mm broad, densely hairy, brown, last 4 segments very narrow and golden colour hairs are present.

Colour

Black	: Eyes.
Blackish brown	: Head, thorax.
Yellowish	: Proboscis, labium, wing, legs.
Brownish	: Antenna, halter.
Reddish brown	: Abdomen.
Male	: 3.20 mm long, slender, smaller than female; antenna plumose, phytophagus.
Host	: Human and cattle.
Host plant	: Unknown.
Holotype	: Female, India, Maharashtra, Kolhapur coll. Jagtap, M.B., 14-VI-2009

	head, antenna, hind leg, wing and abdomen mounted and pinned, labelled as above.
Paratype	: 38 ♂, 123 ♀, sex ratio (M:F) 1:3.23 coll. Jagtap, M.B. Mar. 2006 to Dec. 2009.
Etymology	: The species name *Anopheles kolhapuri* sp. nov. refers to the collection site of mosquitoes *i.e.* Kolhapur city, district Kolhapur, Maharashtra, India.

Distributional Record

3 ♂ 7 ♀ Saswad, 12-III-2006; 5♂ 11 ♀ Vita, 10-VI-2006; 6 ♂ 15 ♀ Patan, 21-VI-2007, 2♂ 9 ♀ Malakapur, 12-VII-2007; 1 ♂ 5 ♀ Pune, 25-VI-2007; 3 ♂ 15 ♀ Junner, 14-VI-2008; 2 ♂, 7 ♀ Medha 20-VII-2008; 1 ♂, 9 ♀, Mhaswad, 23-VIII-2008; 1 ♂ 3 ♀ Kagal, 14-III-2009; 2 ♂ 5 ♀ Wai, 12-IV-2009; 2 ♂ 10 ♀ Satara, 27-VII-2009; 6 ♂ 15 ♀ Ajara, 25-VII-2009; 4 ♂ 12 ♀ Kolhapur, 13-IX-2009.

Remark

According to the key of Rao (1984) this species runs close to *Anopheles* (Anopheles) *gigas* Giles 1901 by following characters:

1. Middle femur having large pale spot.

2. No 6 vein with pale spots on outer half

However, it differs from the above species by having following characters:

1. Dark pale band on forewing is 0.9 mm long and vein no. 3 is dark.

2. Last four segments of abdomen are very narrow and golden colour hairs are present.

3. Scutellum globular and opaque, 0.5 mm long and 0.46 mm broad.

4. White strip on thorax by having longitudinal black and lateral line on thorax. Rest of the thoracic scales are whitish.

5. Flagellar formula:

   $1 \text{ L/W} = 4$, $13 \text{ L/W} = 2.4$, $\text{L1/13} = 1$, $\text{W1/13} = 0.6$, $A = 4$.

Key to the Genera of Tribe Culicini

Proboscis more flexible, usually of uniform thickness but sometimes swollen at tip, not hooked; posterior margin of scutellum more or less trilobed and with three distinct groups of bristles; clypeus longer than broad, rounded above and in front; no v shaped thickening in hind margin of wing.................................... *Culicini*

Key to the Genera Culex and Aedes

1. Margin of squama fringed, vein 6 ending well beyond level of fork of vein 5, Pulvilli present bucco-pharyngeal armature present in female .... *Culex*

2. Margin of squama quite bare, vein 6 short usually ending at about level of fork of vein 5, Proboscis fairly slender and straight; ornamentation and scaling very various .... *Aedes*

Genus *Aedes* Meigen, 1818

Meigen 1818 : 1. Knight and Hull 1951b : 211. Senevet and Andarelli 1954b : 310 (Larvalkey, North Africa). Mattingly 1958 : 1 (Subgenera *Paraedes, Rhinoskusea* and *Cancraedes* Indomalayan area). Smith 1958 : 39 (Female, New England). Sazanova 1958 : 741 (Female, Forest zone, U.S.S.R.). Mattingly 1959 : 1 (subgenera *Skusea, Diceromyia, Geoskusea* and *Christophersimyia, Indomalayan* area).

The genus *Aedes* is erected by Meigen in 1818. The genus *Aedes* contain 600 species represented from all over the world (Cheng, 1964) of which more than 112 species have been reported from India (Barraud, 1934). This genus is characterized by

1. Claws toothed in female.

2. Post spiracular bristles present.

3. Fringe of hair on squma of wing.

4. Female abdomen more pointed.

5. Mesonotum bristles well developed.

6. Pulvilli absent or hair like.

7. Antenna of male always distinctly plumose with the last two segments elongate and in female with moderate long hairs.

8. Cerci longer.

9. Proboscis not enlarged or approximated behind the head.

10. Pronotum without setae

11. Scutellum trilobed.

The genus *Aedes* is divided into following 11 subgenera by Barraud (1934).

1. *Aedes* Meigen, 1818,
2. *Ochlerotatus* Lynch Arribalzaga, 1891,
3. *Medicus* Theobald, 1901,
4. *Stegomyia* Theobald, 1901,
5. *Finlaya* Theobald, 1903,
6. *Bankinella* Theobald, 1907,
7. *Diceromyia* Theobald, 1911,
8. *Christophersiomyia* Barraud, 1923,
9. *Rhinoskusea* Edwards, 1929,
10. *Cancredes* Edwards, 1929,
11. *Indusius* Barrauds, 1934

Barraud (1934) described 116 species under above eleven subgenera from India. Sathe and Girhe (2002) reported 15 species from Maharashtra. Sathe and Tingare (2010) described 21 species out of which 17 were new to the science.

Key to Subgenera of *Aedes*

1. Segment VIII narroal and completely retractile cerci long and narrow and projecting from ring of segment VII .... *Mucidus*

   Segment VIII broader and not completely retractile, cerci shorter and broader 2

2. Tarsal claws toothed at least on fore and mid legs .... *Finlaya*

   Tarsal claw all simple .... *Stegomyia*

Subgenus *Stegomyia* Theobald

Theobald 1901a (June-1), in Howard 1901 : 235, Theobald 1901b : 235 (July 15), 1901 c : ii (Sept), 1901a :

283 (Nov. 23). Logotype : _Culex fosciatus fabricius_ : (Neveu-Lemaire 1902 : 211).

Scutomyia Theobald 1904 b : 77 Haplotype : _albolineata_ Theobald.

Mattingly 1952 : 235 (Ethiopian Region). Mattingly 1953 : 1 (Ethiopian Region).

The subgenus _Stegomyia_ is characterized by the following features:

1. Size of mosquito medium or small.
2. Highly ornamented body with colour black or dark and patches spots or lines of snow white scales.
3. Two or more basal white bands on tarsi or at least one pair of leg of which one or more tarsal segments completely white.
4. In few cases proboscis entirely dark.
5. Harpago absent.
6. Antenna of male with plumose hairs mainly projecting on two side.
7. Broad and flat scales on vertex and scutellum.
8. Palpi of male slender with few hairs, terminal segments unruptured.

Barraud described 16 species under _Stegomyia_, from India. Seven species from Kolhapur district have been reported under the subgenus _Stegomyia_ of which two are redescribed and four species have been newly reported and described for the first time from India (Sathe and Girhe, 2002).

M.C.I., p. 233, Genotype, _Culex fasciatus_ Fab. _Scutomyia_ Theobald, 1904, Entom. xxxvii, p. 77, Genotype _S. albolineata_

Theo. *Quasistegomyia* Theobald, 1906, 2nd Rept. Wellac. lab. p. 69. Genotype *Q. unilineata* Theo. *Catatassomyia* Dyar and Shannon, 1925, Insec. Ins. Mens. xiii, p.71, Genotype, *C. neronephada* D. and S.

Key to the Species of Subgenera *Stegomyia*

1. All tibiae with white rings 2

   Tibiae without white rings 3

2. Mesonotum marked with 4-6 small white spots; femora with preapical white rings; proboscis with scattered yellowing scaling .... *vittatus*

   Mesonotum otherwise marked; femora without preapical white rings, proboscis entirely dark... *desmotes*

3. Mesonotum marked with a pair of lateral curved white lines, and usually also with a pair of submedian yellowish lines; two dots of white scales on clypeus .... *aegypti*

   Mesonotum otherwise marked; clypeus without white scales 4

4. Palpi entirely dark, proboscis thin and longer than fore femur; mesonotum with a fairly narrow, silvery, median stripe in front; last two hind tarsal segments all dark ....*albolineatus*

   Palpi with white scaling, proboscis normally thick and about length of fore femur 5

5. Mesonotum with a narrow median white line running nearly the whole length 6

Mesonotum with a broad white anterior
stripe, or with a white patch or pair of
patches in front 10

6. Mid-femur with a preapical white spot
on anterior surface .... *unilineatus*
Mid-femur without a white spot 7

7. White scales on pleurae arranged more or
less in two lines; a line of white flat scales
over wing-root continued nearly to lateral
lobe of scutellum .... *scutellaris*
White scales on pleurae arranged in
irregular patches : a patch of white
scales in front of wing-root only 8

8. Pale markings on mesonotum except for
the median white stripe, of a yellowish
colour .... *flavopictus*
Pale markings on mesonotum all silvery white 9

9. Abdomen without silvery bands on
dorsum (? always) .... *subalbopictus*
Abdomen with silvery basal bands .... *albopitcus*
on dorsum .... *pseudalbopictus novalbopictus*

10. Mid-femur with a median white spot
on anterior surface; mesonotum with
several white patches .... *w-albus*
Mid-femur without a white spot 11

11. Mesonotum with a broad median white
stripe, narrowing posteriorly .... *mediopunctatus*
Mesonotum with a roundish white
area in front 12

12. Mesonotum with a small median white
spot in front; all segments of hind tarsi
white-ringed .... *edwardsi*

Mesonotum with a large median white
patch in front; last one or two hind
tarsal segments all dark .... *annandalei craggi*

Aedes (Stegomyia) *aegypti* Linnaeus, 1762

Female (Figure 111)

3.90 mm long, 0.85 mm broad, dark brown, without
ornamentation, with scales on body and legs; head 0.60
mm long, 0.52 mm broad, black; antenna 1.65 mm long,
blackish brown; thorax 1.20 mm long, 0.85 mm broad,
brown; forewing 3.15 mm long, 0.75 mm broad; hind leg
7.60 mm long; abdomen 2.16 mm long, 0.60 mm broad,
blackish brown.

Head (Figure 83)

0.60 mm long, 0.52 mm broad, black, globular, with
flat scales; compound eyes black, large, rounded, ocular
space 0.33 mm; interocular distance 0.25 mm, vertex with
dark silvery flat scales, hairy, with rod like vertical bristles;
two small silvery white dots on clypeus; nape 0.06 mm long,
0.09 mm broad, black, triangular; proboscis 2.16 mm long
and moderately slender, cylindrical, brown scally; labium
1.90 mm long, slender; labellum 0.26 mm long, brown; palpi
short 0.31 mm long, three segmented, shorter than
proboscis, tip of palpi conspicuously white; palpifer brown,
palpus brown, scally; mandibles and maxillae as long as
proboscis, yellow, straight, palpi with scaling, proboscis thick
and about the length of fore femur.

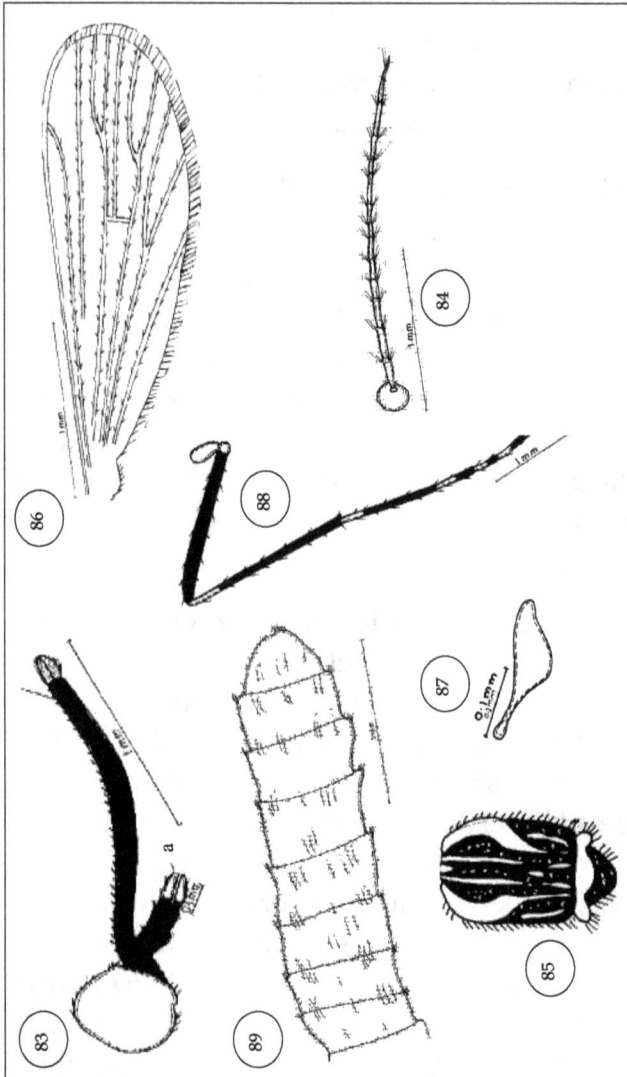

Plate 14–*Aedes* (Stegomyia) *aegypti* sp. nov.

Figure 83: Head-Palpi (a) and Proboscis (b); **Figure 84**: Antenna; **Figure 85**: Thorax; **Figure 86**: Forewing; **Figure 87**: Halter; **Figure 88**: Hind leg; **Figure 89**: Abdomen

Antenna (Figure 84)

1.65 mm long, blackish brown, 15 segmented, pilose, hairy; scape 0.05 mm long, 0.15 mm broad, brown; pedicel 0.25 mm long, 0.10 mm broad, rounded, brown, flagellum 1.35 mm long, 13 segmented.

Flagellar Formula

3 L/W = 4.5, 9 L/W = 5.33, L 3/9 = 1.12, W 3/9 = 1.33, A = 3.07.

Thorax (Figure 85)

1.20 mm long, 0.85 mm broad, golden brown, laterally compressed; scutum brown, posterior pronotal lobe brown; scutellum 0.5 mm long, 0.6 mm broad, flat silvery white scales on all lobes of scutellum, brown, rounded; sternopleuron brown; mesepimeron broad, flat, with white patches; mesothoracic spiracles rounded; pronotum lack of satae. Thorax with conspicuous and well defined ornamentation in the form of either of lines or broad patches of silvery white scales. Half circular white mark and some white spots are present on thorax.

Forewing (Figure 86)

3.15 mm long, 0.75 mm broad, unspotted, dark scales on squama and alula, scales present along the veins as well as hind margin; subcosta straight, 2.70 mm long, reaching the costa; media straight, bifurcated apically; radius 2.10 mm long, straight slightly curved at end; cubitus bifurcated.

Halter (Figure 87)

0.43 mm long, 0.09 mm broad, yellow, rounded in lateral view.

Hind Leg (Figure 88)

7.60 mm long, slender, longer than body; coxa 0.20 mm long, yellowish, broad, with 4 large and 7 small coxal bristles; trochanter 0.15 mm long and 0.10 mm broad, rounded, hard, brown; femur 1.90 mm long, cylindrical; tibia 2.30 mm long dark brown, with white ring; tarsus 3.05 mm long, five segmented, tarsi with narrow basal white bands to first 2 or 3 segments; 1st tarsal segment 1.20 mm long, 2nd tarsal segment 1.05 mm long, 3rd tarsal segment 0.55 mm long, 4th tarsal segment 0.15 mm long, 5th tarsal segment 0.10 mm long, with dark and white bands, 1st hind tarsal segment less than tibia; claw curved, two or more white bands on tarsi, mid femura with preapical white spots.

Other Legs

Special marks : similar.

Abdomen (Figure 89)

2.16 mm long, 0.60 mm broad, blackish with snow white marking and basal bands, slender, pointed, black and white transverse bands, white patches of scales, laterotergites, sternal plate dirty white, segment VIII broader and not completely retractile, sternite VIII large and not very prominent in repose, tergal plates dark blackish brown; post genital plate 0.90 mm long, 0.07 mm broad; anal cerci rather short and broad, 0.02 mm long and 0.03 mm broad, brown, flat, hairy. Front and middle claws of the female either toothed or not.

Colour

Black   : Eyes, head.
Brown   : Proboscis, scape, thorax.
Blackish Brown   : Abdomen.

Yellow	: Mandibles, maxillae, wing.
Male	: 3.10 mm long, slender, smaller than female; antenna plumose brushy, phytophagous.
Host	: Human and Cattle.
Host Plant	: Unknown.
Holotype	: Female, India, Maharashtra Jaysingpur coll. Jagtap, M.B., 12-XII-2009 head, antenna, hind leg, wing and abdomen mounted and pinned, labeled as above.
Paratype	: 38 ♂, 90 ♀, sex ratio (M:F) 1:2.36 coll. Jagtap, M.B. Mar. 2006 to Dec. 2009.
Etymology	: The species name *Aedes* (Stegomyia) *aegypti* (Linnaeus), 1762.

Distributional Record

1 ♂ 3♀ Jath, 5-III-2006; 3 ♂ 4 ♀ Kolhapur, 24-VI-2006; 1 ♂ 4 ♀ Jaysingpur, 9-VII-2006; 2 ♂ 5 ♀ Miraj, 11-VI-2006; 2 ♂ 7 ♀ Kolhapur 24-VI-2006, 3 ♂ 8 ♀ Kagal 12-VIII-2006; 3 ♂ 6 ♀ Malakapur, 11-II-2007; 4 ♂, 8 ♀, Pune, 25-VII-2007; 3 ♂ 8 ♀ Satara, 23-XI-2007; 1 ♂, 4 ♀, Wai, 9-II-2008; 0 ♂ 5 ♀ Ajara, 26-IV-2008; 2 ♂ 2 ♀ Kagal, 12-VII-2008; 4 ♂ 12 ♀ Mhaswad, 23-VIII-2008; 4 ♂, 14 ♀, Kolhapur, 14-VI-2009; 5 ♂, 15♀, Jaysingpur 12-XII-2009.

Remark

According to the key of Barraud (1934) this species runs close to *Aedes aegypti* by having following characters.

1. Thorax with conspicuous and well defined ornamentation in the form of either of lines or broad

patches of silvery white scales. Half circular white mark and some white spots are present on thorax.

2. Two small silvery white dots on clypeus.

3. Eight segment of abdomen is large but distinctly retractile.

However, some following additional characters have been observed:

1. Scutellum shape (rounded) and size (0.5 mm long and 0.6 mm broad).

2. Halter shape (rounded) and size (0.45 mm long and 0.09 mm broad).

3. Flagellar formula :
   $3 L/W = 4.5, 9 L/W = 5.33, L 3/9 = 1.12, W 3/9 = 1.33, A = 3.07$.

4. Phyllogenetically it runs close to *Aedes aegypti*. However, it differs from 14 species by having NHJ branch length = 7.35313831.

Aedes (Stegomyia) *albopictus* Skuse, 1894

Female (Figure 112)

3.75 mm long, 0.75 mm broad, dark brown, with ornamentation, with scales on body and legs; head 0.60 mm long, 0.56 mm broad, black; antenna 1.70 mm long, blackish brown; thorax 1.15 mm long, 0.75 mm broad, brown; forewing 3.09 mm long, 0.85 mm broad; hind leg 8.40 mm long; abdomen 2.10 mm long, 0.65 mm broad, blackish brown. It easily recognized by its black and white coloration.

Head (Figure 90)

0.60 mm long, 0.56 mm broad, black, globular, with flat scales; compound eyes black, large, rounded, ocular space 0.33 mm; interocular distance 0.25 mm, vertex dark black, hairy, with rod like vertical bristles; clypeus without white scales; nape 0.06 mm long, 0.09 mm broad, black, triangular; proboscis 2.15 mm long, cylindrical, brown scally; labium 1.90 mm long, slender; labellum 0.25 mm long, brown; palpi 0.35 mm long, three segmented, shorter than proboscis; brown, palpus brown, scally; mandibles and maxillae as long as proboscis, yellow, straight, palpi with scaling, proboscis thick and about the length of fore femur.

Antenna (Figure 91)

1.75 mm long, blackish brown, 15 segmented, pilose, hairy; scape 0.05 mm long, 0.15 mm broad, brown; pedicel 0.25 mm long, 0.10 mm broad, rounded, brown, flagellum 1.45 mm long, 13 segmented.

Flagellar Formula

3 L/W = 4.75, 9 L/W = 6.33, L 3/9 = 1.11, W 3/9 = 1.33, A = 4.38.

Thorax (Figure 92)

1.15 mm long, 0.75 mm broad, golden brown, laterally compressed; scutum brown, posterior pronotal lobe brown; scutellum 0.5 mm long, 0.6 mm broad, brown, rounded, with silvery white scales; sternopleuron brown; mesepimeron broad, flat, with white patches; mesothoracic spiracles rounded; pronotum lack of satae. The dorsal surface of the thorax is ornamented by a single white strip along the median line running nearly whole length of mesonotum.

Forewing (Figure 93)

3.09 mm long, 0.85 mm broad, unspotted, dark scales on squama and alula, scales present along the veins as well as hind margin; subcosta straight, 2.75 mm long, reaching

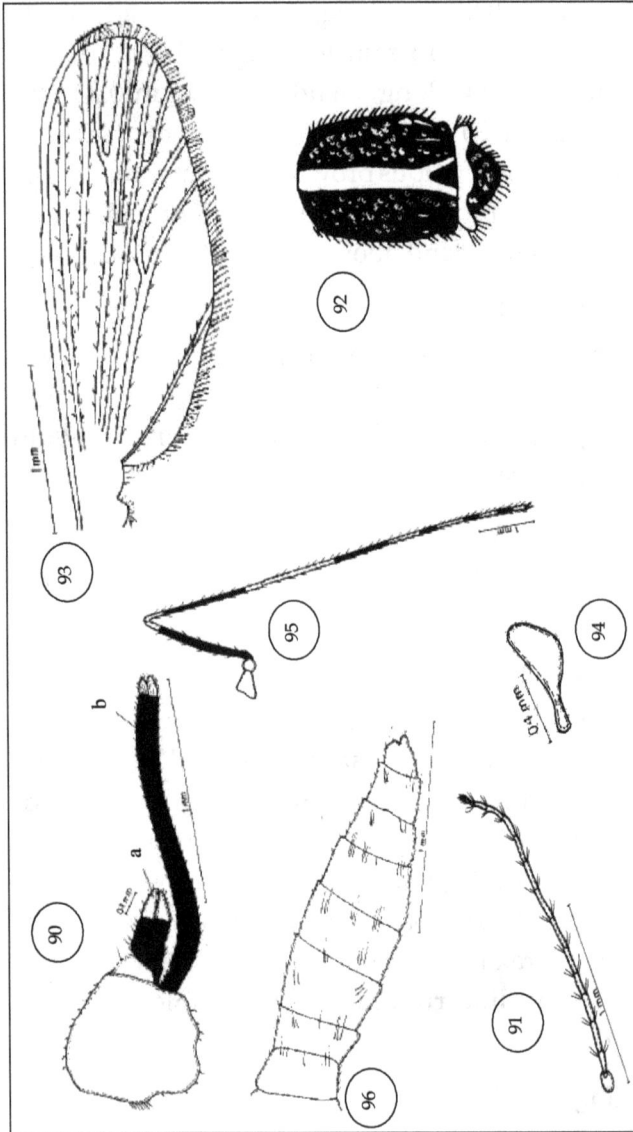

Plate 15—*Aedes* (Stegomyia) *albopictus* sp. nov.

Figure 90: Head-Palpi (a) and Proboscis (b); **Figure 91**: Antenna; **Figure 92**: Thorax; **Figure 93**: Forewing;
Figure 94: Halter; **Figure 95**: Hind leg; **Figure 96**: Abdomen

the costa; media straight, bifurcated apically; radius 2.15 mm long, straight slightly curved at end; cubitus bifurcated.

Halter (Figure 94)

0.40 mm long, 0.08 mm broad, yellow, rounded in lateral view.

Hind Leg (Figure 95)

8.10 mm long, slender, longer than body; coxa 0.25mm long, yellowish, broad, with 4 large and 7 small coxal bristles; trochanter 0.15 mm long and 0.10 mm broad, rounded, hard, brown; femur 2.00 mm long, cylindrical; tibia 2.50 mm dark brown, with white ring; tarsus 3.20 mm long, five segmented; 1st tarsal segment 1.25 mm long, 2nd tarsal segment 1.05 mm long, 3rd tarsal segment 0.60 mm long, 4th tarsal segment 0.20 mm long, 5th tarsal segment 0.10 mm long, with dark and white bands, 1st hind tarsal segment less than tibia; claw curved, broad white rings on all segments of hind tarsi.

Other Legs

Special marks : similar.

Abdomen (Figure 96)

2.10 mm long, 0.65 mm broad, blackish with snow white marking and basal bands, slender, pointed, black and white transverse bands, white patches of scales, laterotergites, sternal plate dirty white, segment VIII broader and not completely retractile, sternite VIII large, tergal plates dark blackish brown; post genital plate 0.90 mm long, 0.07 mm broad; anal cerci short and broad, 0.02 mm long and 0.03 mm broad, brown, flat, hairy.

Mosquito Diversity and Control

Colour

Black	: Eyes, head.
Brown	: Proboscis, scape, thorax.
Blackish Brown	: Abdomen.
Yellow	: Mandibles, maxillae, wing.
Male	: 3.25 mm long, slender, smaller than female; antenna plumose brushy, phytophagous.
Host	: Human and Cattle.
Host Plant	: Unknown.
Holotype	: Female, India, Maharashtra Kolhapur, coll. Jagtap, M.B., 14-VI-2009 head, antenna, hind leg, wing and abdomen mounted and pinned, labeled as above.
Paratype	: 35 ♂, 101 ♀, sex ratio (M:F) 1:2.88 coll. Jagtap, M.B. Mar. 2006 to Dec. 2009.

Distributional Record

1 ♂ 3 ♀ Kolhapur, 24-VI-2006; 0 ♂ 4 ♀ Jaysingpur, 9-VII-2006; 2 ♂ 5 ♀ Miraj, 11-VI-2006; 3 ♂ 7 ♀ Kolhapur 24-VI-2006, 4 ♂ 9 ♀ Kagal 12-VIII-2006; 1 ♂ 3 ♀ Malakapur, 11-II-2007; 2 ♂ 4 ♀ Koregaon, 3-V-2007; 1 ♂ 3 ♀ Junner, 16-VIII-2007; 3 ♂ 11 ♀ Satara, 23-XI-2007; 3♂, 7 ♀, Wai, 9-II-2008; 1 ♂, 3 ♀, Vita, 23-III-2008; 3 ♂ 8 ♀ Kagal, 12-VII-2008; 0 ♂ 3♀ Medha, 20-VII-2008; 0 ♂, 2 ♀, Bhor, 10-VIII-2008; 2 ♂ 6 ♀ Mhaswad, 23-VIII-2008; 4 ♂, 9 ♀, Kolhapur, 14-VI-2009; 2 ♂ 6 ♀ Patan, 11-VII-2009; 3 ♂, 8 ♀, Jaysingpur 12-XII-2009.

Remark

According to the key of Barraud (1934) this species runs close to *Aedes albopictus* by having following characters.

1. Body recognized by black and white coloration.
2. The dorsal surface of the thorax is ornamented by a single white strip along the median line running nearly whole length of mesonotum.
3. Broad white rings on all segments of hind tarsi.

However, some following additional characters have been observed:

1. Scutellum shape (rounded) and size (0.5 mm long and 0.6 mm broad).
2. Halter shape (rounded) and size (0.40 mm long and 0.08 mm broad).
3. Flagellar formula :
   $3 L/W = 4.75$, $9 L/W = 6.33$, $L 3/9 = 1.11$, $W 3/9 = 1.33$, $A = 4.38$.
4. Phyllogenetically it runs close to *Aedes albopictus*. However, it differs from 17 species by having NHJ branch length = 6.40278786.

Subgenus *Mucidus* Theobald, 1901

Theobald 1901 b : 235 (July 15); 1901 c : 2 (Sep.). 1901a : 268 (Nov. 23). Logotype : *Culex alternans* Westwood (Neveu-Lemaire 1902 : 219). Knight 1947 : 315. Mattingly 1961 : 17 (Indomalayan area). Tyson 1970 : 28. (rev. Southeast Asia).

M.C., i, p. 268. Genotype *Culex alternans* Westwd.

Pardomyia Theobald, 1907, M.C. iv, p. 280.

Genotype, *P. aurantia,* Theo.

Elcrinomyia leicester, 1908, Cul. Malaya, p. 71.

Genotype, *E. aurcostriata* Leic.

The subgenus *Mucidus* Theobald is distinguished from other subgenera by having following characters :

1. Adults large sized with outstanding yellow, white and brown scales on body and legs, giving them mottled and mouldy appearance.
2. Wing mottled with yellow, brown and creamy scales.
3. Mesonotum remarkably long and twisted, resembling strands of cotton-wool.
4. ppn bristles numerous, usually about 20.

From India only two species have been described under this subgenus. From Kolhapur district a new species, *Aedes indica* sp. nov. have been reported and described for the first time.

Key to the Species of Subgenus *Mucidus* Theobald

1. Palpi more than ½ length of proboscis .. *scatophagoides*

   palpi short, not more than ½

   length of proboscis legs blackish..

   with whitish markings, mesonotal

   scales not brown .... *tasgaonensis* sp. nov.

   Legs yellowish brown with white

   Bands, big size mosquito,

   5 white bands on tarsi ....*sathei* sp. nov.

Aedes (Mucidus) *sathei* sp.nov.

Female (Figure 113)

8.64 mm long, 1.90 mm broad, big size mosquito with dark brown colour, without ornamentation, with scales on body and legs; head 1.1 mm long, 0.90 mm broad, brown; antenna 2.70 mm long, yellowish brown; thorax 2.40 mm long, 1.90 mm broad, brown; forewing 6.30 mm long, 2.64 mm broad; hind leg 17.40 mm long; abdomen 5.10 mm long, 0.80 mm broad, yellowish brown.

Head (Figure 97)

1.1 mm long, 0.9 mm broad, brown, globular, with flat scales; compound eyes black, large, rounded, ocular space 0.17 mm, interocular distance 0.15 mm; vertex brownish, hairy, with rod like vertical bristles; clypeus without white scales; proboscis (Figure 1b) 3.70 mm long, cylindrical, brown scally; labium 3.10 mm long, slender; labellum 0.60 mm long, brown; palpus 2.10 mm long, more than half of proboscis, three segmented, shorter than proboscis; palpifer brown, palpus brown, scally; yellow, straight, palpi with scaling, proboscis long and thick and about the length of fore femur. Head with white stripe on either side of middle line or not dorsally dark.

Antenna (Figure 98)

2.70 mm long, brownish, 14 segmented, pilose, hairy; scape 0.09 mm long, 0.35 mm broad, brown; pedicel 0.47 mm long, 0.42 mm broad, rounded, brown; flagellum 2.23 mm long, 13 segmented.

Flagellar Formula

3 L/W = 3.2, 9 L/W = 4, L 3/9 = 0.8, W 3/9 = 1, A = 2.25.

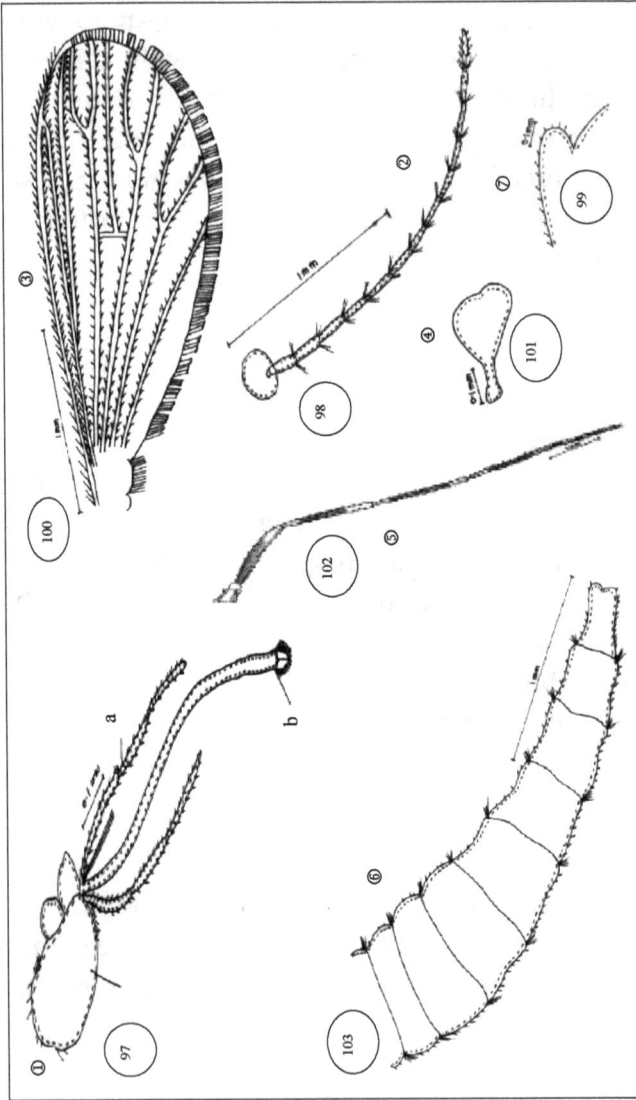

Plate 16—Aedes (Mucidus) *sathei* sp. nov.

Figure 97: Head, Palpi (a) and Proboscis (b); **Figure 98**: Antenna; **Figure 99**: Scutellum; **Figure 100**: Forewing; **Figure 101**: Halter; **Figure 102**: Hindled; **Figure 103**: Abdomen

Thorax

2.44 mm long, 1.90 mm broad, brown, laterally compressed; scutum brown, posterior pronotal lobe brown; scutellum (Figure 99) 0.40 mm long, 0.24 mm broad, brown, rounded, with long brown scales; sternopleuron brown; mesepimeron broad, flat, with pale patches; mesothoracic spiracles rounded; pronotum lack of satae.

Forewing (Figure 100)

6.30 mm long, 2.64 mm broad, unspotted, dark scales on squama and alula, scales present along the veins as well as hind margin; subcosta straight, 4.75 mm long, reaching the costa; media straight, bifurcated apically; radius 5.37 mm long, straight slightly curved at end; cubitus bifurcated.

Halter (Figure 101)

0.48 mm long, 0.15 mm broad, yellow, heart shaped.

Hind leg (Figure 102)

17.40 mm long, slender, longer than body; coxa 0.45 mm long, yellowish brown, broad; trochanter 0.30 mm long, rounded, hard, brown; femur 3.50 mm long, cylindrical; tibia 3.90 mm long dark brown; tarsus 9.25 mm long, five segmented; 1st tarsal segment 3.75 mm long, 2nd tarsal segment 2.45 mm long, 3rd tarsal segment 1.40 mm long, 4th tarsal segment 0.90 mm long, 5th tarsal segment 0.75 mm long with dark and pale bands, 1st hind tarsal segment broad but less than tibia; claw toothed and curved, basal white bands with five white bands on tarsi, mid femura with preapical white spots.

Other Legs

Special marks : similar.

Abdomen (Figure 103)

5.10 mm long, 0.80 mm broad, blackish with snow white marking and basal bands, slender, pointed, brown and with pale transverse bands; white patches of scales, laterotergites, sternal plate dirty white; segment VIII broader and not completely retractile, sternite VIII large, tergal plates dark brown; post genital plate 0.8 mm long, 0.4 mm broad; anal cerci long and narrow, 0.20 mm long, brown, flat, hairy.

Colour

Black	: Eyes.
Brown	: Proboscis, head, legs, scape, thorax, abdomen.
Yellow	: Mandibles, maxillae, wing.
Male	: 6.20 mm long, slender, smaller than female; antenna plumose, brushy, phytophagous.
Host	: Human beings, Cattle.
Host Plant	: Unknown.
Holotype	: Female, India, Maharashtra, Rukadi coll. Jagtap, M. B., 18-IX-2009 bead, antenna, hind legs, abdomen, mounted and pin up.
Paratype	: 1 ♂, 3 ♀ sex ratio (M:F) 1:3. coll. Jagtap, M. B. Jan 2007 to Dec. 2009.
Etymology	: The species name *Aedes* (Mucidus) *sathei* sp. nov is the honour to my research guide Prof. T. V. Sathe.

Distributional Record

0 ♂, 1 ♀ Miraj 14-III-2007; 1 ♂, 1 ♀ Rukadi 15-IX-2009; 0 ♂, 1♀, Jaysingpur 12-XII-2009.

Remarks

According to the key of Barraud (1934) this species runs close to *Aedes* (Mucidus) *scatophagoides* by having following characters:

1. Palpi more than one half length of proboscis.
2. Legs with whitish marking.

However, it differs by having following characters:

1. Legs not blackish brown but yellowish brown with white bands.
2. White stripes transversely from middle white line on the head.
3. Big size mosquito with brownish colour.
4. Five white bands on tarsi.
5. Scutellum rounded with long brown scales (0.40 mm long 0.24 mm broad).
6. Rod like vertical bristles on vertex.
7. Halter heart shaped.
8. Flagellar formula:

   3 L/W = 3.2, 9 L/W = 4, L 3/9 = 0.8, W 3/9 = 1, A = 2.25.

Subgenus *Finalaya* Thebald 1903

Theobald 1903a : 281, Logotype : *Culex kochi* Doenitz (Blanchard. 1905 : 415), Giles 1904a : 366 (As *Finlaya* emend.), *Danielsia* Theobald 1904b : 78. Haplotype : *albotaeniata* Leicester.

Hulecoeteomyia Theobald 1904 : 163, Haplotype : *trilineata* Leicester. Popea Ludlow 1905 : 95. Haplotype : *Lutea* Ludlow.

M.C. iii, p. 281, Genotype, *F. poicilia* Theo.

Finlaya Giles (emend.), 1904, *Journ. Trop. Med.* vii, p. 366.

Danielsia Theobald, 1904, *Entom.* xxxvii, p. 78, Genotype, *D. albotaeniata* Theobald. *Hulecaetcomyia* Theobald, 1904, *Entom.* xxxvii, p. 162, Genotype, *H. trilineata* Theo.

Popea Ludlow, 1905, Can. Ent. xxxvii, p. 95, Genotype, *P. lutea* Ludlow. *Phagomyia* Theobald, 1905, Gen. Insect., Fam. Culicidae, p. 21. Genotype, *P. gubernatoris* (Giles).

Lepidotomyia id., ib. p. 22 Genotype, *L. magna* Theo.

Pseudocarrollia Theobald, 1910, Rec. Ind. Mus. iv, p. 13, Genotype, *P. lophoventralis* Theo.

From India 40 species have been described. This sub genus is widely distributed in India.

Key to the Species of Subgenus *Finalaya*

1. Wings elaborately spotted and speckled with black and white scales .... *poecilus*

   Wings not spotted or speckled, wing-scales all dark, except for a short line of pale one at base of costa in some species 2

2. Tarsi entirely dark.. 3

   Tarsi with white markings 7

3. Mid-femur with a median silvery mark on anterior surface .... *dissimilis*

   Mid-femur without such mark 4

4. Mesonotum with a large snowy-white patch
   in front, which may, in the ♀, be more or
   less divided into lateral patches .... *niveus* group

   Mesonotum with ochreous, yellow, or
   golden scales 5

5. Abdominal sternites with orange
   patches .... *pulchriventer*

   Abdominal sternites without orange
   patches 6

6. Mesonotum, in ♀, black, with ochreous
   scales arranged in lines; in ♂ entirely,
   but sparsely, covered with pale scales .... *oreophilus*

   Mesonotum covered with ochreous
   scales, with a pair of indistinct
   submedian dark lines .... *suffusus*

7. Hind tarsi with one or more white rings
   at bases of segments only 8

   Hind tarsal segments with both apical
   and basal white rings 23

8. Venter of abdomen with orange
   markings .... *auronitens*

   Venter of abdomen without orange
   markings 9

9. Hind tarsi with only one white ring .... *unicinctus*

   Hind tarsi with three or four white rings 10

10. Hind tarsi with three white rings 11

    Hind tarsi with four white rings 20

11. Mesonotum marked with narrow lines
    of golden scales 12

    Mesonotum marked otherwise 16

12. Proboscis with a pale ring or with pale
    scaling on underside 13

    Proboscis entirely dark................. *saxicola*

13. Proboscis with pale scaling both on upper
    and undersides 14

    Proboscis pale on underside only 15

14. Proboscis pale on basal 4/5 both on upper
    and undersides .... *pallirostris*,♀,

    Proboscis extensively pale on underside
    and with white scaling forming a
    narrow band on upper side .... *chrysolineatus*

15. Mid-femur dark on anterior aspect ....*harveyi*

    Mid-femur with a pale line on
    basal ½ anteriorly..............................*formosensis*

16. Mesonotum blackish, with a white
    spot in front .... *stevensoni*

    Mesonotum otherwise 17

17. All femora with small white knee-spots,
    scales on dorsum of head mainly
    narrow ....*christophersi*

    All femora dark at knee, scales on dorsum
    of head mainly broad and flat.................... 18

18. Posterior, or under, surface of fore tibia
    conspicuously pale for whole length ....*gilli*

    Fore tibia dark, except narrowly at base 19

19. Mid-femur dark on anterior surface,
    except at extreme base .... *simlensis*

    Mid-femur with a well-defined white
    streak on anterior surface, ventrally, on
    basal ½ .... *albocinctus*

20. Hind tarsi with white ring on segment
    4 very wide and covering nearly the
    whole segment .... *subsimilis*

    5 white rings on hind tarsi, abdominal
    venter, dark brown, mid tibia pale at
    posterior .... *rajashri* sp. nov.

    Hind tarsi with white ring on segment
    4 not very wide 21

21. Mesonotum with a large area of white
    scaling in front ....*albotoeniatus*

    Mesonotum dark, or with pale scaling
    forming lines 22

22. Mesonotum dark on anterior ½, or with
    an indistinct median yellow line ....*albotoeniatus*
    .... var. *miki*

    Mesonotum with a median and sublateral
    lines of white or creamy scales .... *shortti*

23. Mesonotum with a white patch in front 24

    Mesonotum without a white patch in front 33

24. Venter of abdomen with very long outstanding
    tufts of scales 25

    Venter of abdomen with only moderately
    developed tufts of outstanding scales or
    with none 26

25. ppn bare .... *khazani*

ppn with white scales .... *prominens*

26. ppn with only a small patch of of
white scales on posterior border 27

ppn with a large patch of white scales..................... 29

27. Venter of abdomen with moderately
developed tufts of outstanding scales 28

Venter of abdomen without tufts of
outstanding scales.................................... *cogilli*

28. Fore tibia dark on posterior surface, except
for an apical white ring; scutellum densely
clothed with flat white scales; fore femur
dark on upper, or dorsal, surface
at apex .... *lophoventrulis*

Fore tibia pale posteriorly for whole
length; scutellum much less densely
clothed with white scales; fore femur
with a white spot at apex on upper,
or dorsal, surface .... *cacharanus*

29. Scutellar scales brownish-black 30

Scutellar scales mainly white 32

30. Head without a median white-scaled
line in ♀..*feegradei*

Head with a median white-scaled line 31

31. Abdomen with small outstanding tufts
of scales on venter and roughened scales
on dorsum .... *assamensis*, ♀

Abdomen without obvious tufts of
outstanding scales.................................*deccanus*

32. Venter of abdomen with small outstanding
    tufts of scales; mesonotum almost entirely
    covered with white scales .... *assamensis*, ♀

    Venter of abdomen without tufts
    of outstanding scales; white scales
    of mesonotum confined to an
    anterior patch and one in front
    of each wing-root .... *gubernatoris*

33. Mesonotum marked with lines of
    white, yellowish, or golden scales
    on a dark ground 34

    Mesonotum covered with ochreous,
    brown, or golden scales, not arranged in lines 38

34. Proboscis entirely dark 35

    Proboscis with pale scaling on undersurface 37

35. Mesonotum with a line of creamy scales
    each side, continued over wing-root, no
    median line, but three small patches of
    pale scales on anterior margin .... *sintoni*

    Mesonotum with median and lateral
    well-defined lines of white or golden scales;
    no patch of scales on anterior border of
    mesonotum 36

36. Pale lines on mesonotum white or creamy;
    femora with white longitudinal lines for
    whole length .... *pseudotoeniatus*

    Pale lines on mesonotum golden,
    femora not lined with white .... *greeni* var. *kanaranus*

37. Proboscis pale on underside for basal 2/3, pale lines on mesonotum very narrow, clearly defined and golden .... *macdougalli*

    Proboscis pale beneath from near base to tip, except for a small interruption at about ¾ from base; lines on mesonotum pale yellow, the median one framed of two lines of scales placed close together ....*elsiae*

38. Last segment of hind tarsi white dorsally; mesonotal scales bright golden .... *greeni*

    Last segment of hind tarsi dark; mesonotal scales brown in ♀ pale yellow in ♂. ....*inquinatus*

Aedes (Finalaya) *rajashri* sp.nov.

Female (Figure 114)

4.05 mm long, 1.30 mm broad, dark brown, no shaggy appearance, without ornamentation, with scales on body and legs; head 0.70 mm long, 0.50 mm broad, black; antenna 1.65 mm long, blackish brown; thorax 1.10 mm long, 1.30 mm broad, brown; forewing 2.70 mm long, 0.65 mm broad; hind leg 7.80 mm long; abdomen 2.25 mm long, 0.55 mm broad, blackish brown.

Head (Figure 104)

0.70 mm long, 0.50 mm broad, black, globular with white spots, with flat scales; compound eyes black, large, rounded, ocular space 0.35 mm; interocular distance 0.23 mm, median white scale line absent, vertex dark black with two white spots, hairy, with rod like vertical bristles; clypeus

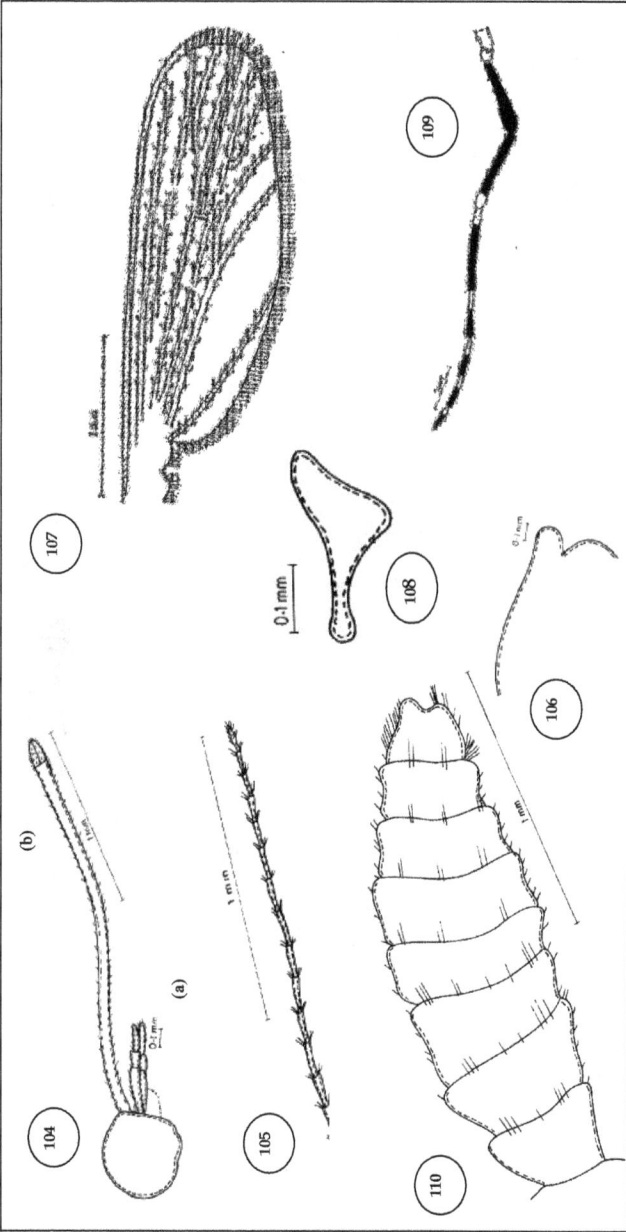

Plate 17–*Aedes* (Finalaya) *rajashri* sp. nov.

Figure 104: Head-Palpi (a) and Proboscis (b); **Figure 105**: Antenna; **Figure 106**: Scutellum; **Figure 107**: Forewing; **Figure 108**: Halter; **Figure 109**: Hind leg; **Figure 110**: Abdomen

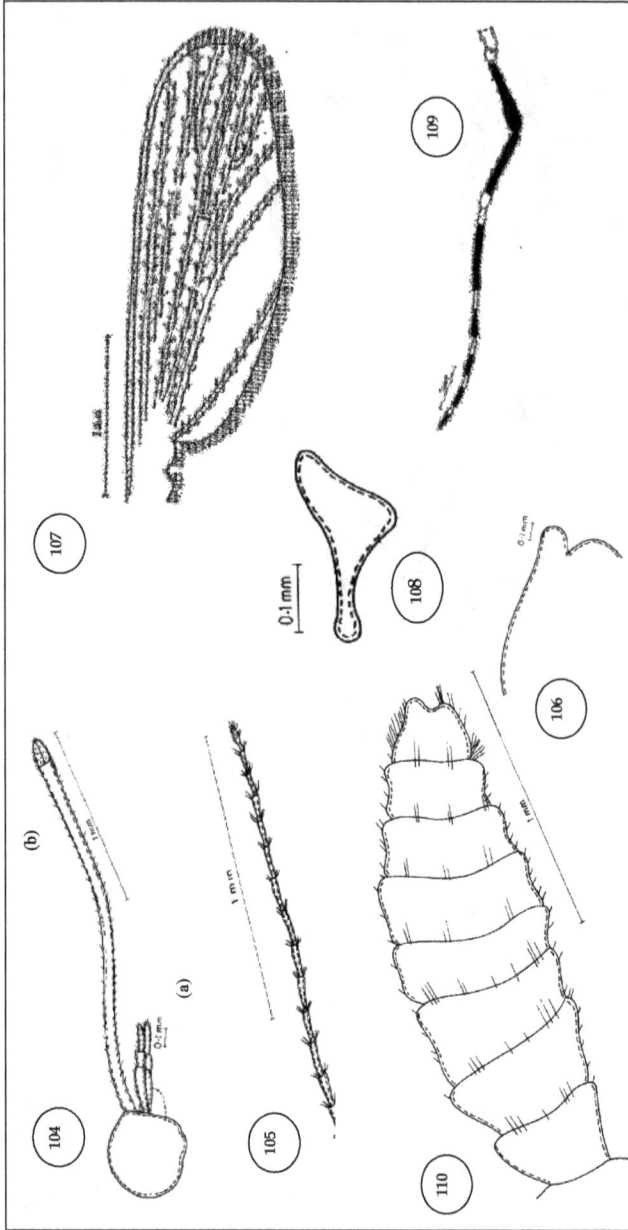

Plate 17–*Aedes* (Finalaya) *rajashri* sp. nov.

Figure 104: Head-Palpi (a) and Proboscis (b); **Figure 105**: Antenna; **Figure 106**: Scutellum; **Figure 107**: Forewing; **Figure 108**: Halter; **Figure 109**: Hind leg; **Figure 110**: Abdomen

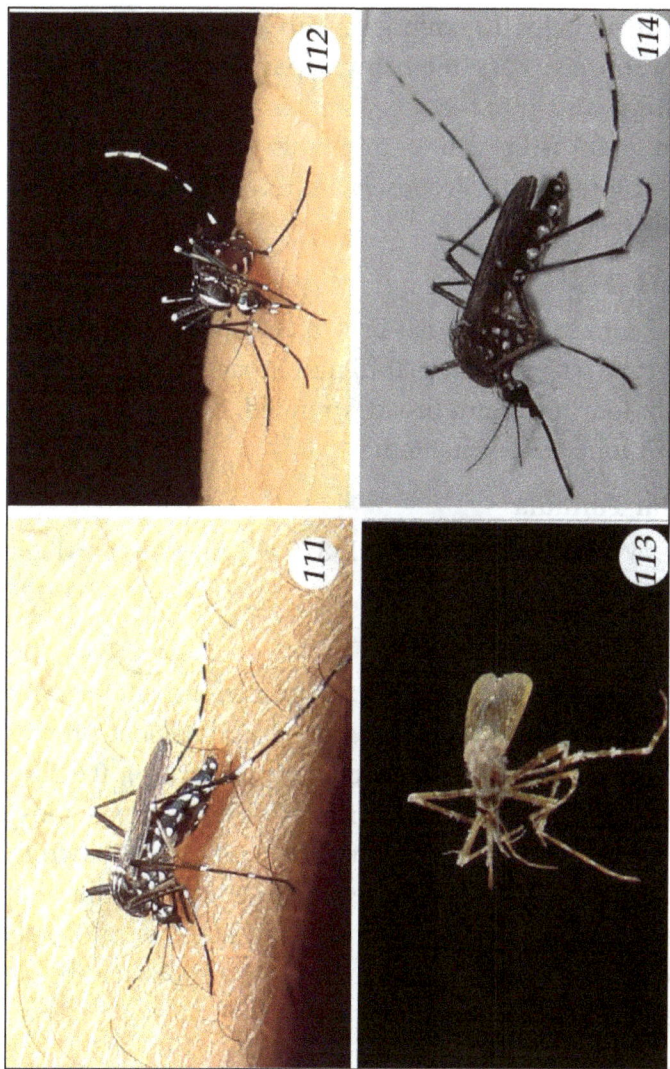

Plate 18—Figure 111: *Aedes* (Stegomyia) *aegypti* sp. nov.; **Figure 112**: *Aedes* (Stegomyia) *albopictus* sp. nov.; **Figure 113**: *Aedes* (Mucidus) *sathei* sp. nov.; **Figure 114**: *Aedes* (Finalaya) *rajashri* sp. nov.

without white scales; nape 0.05 mm long, 0.09 mm broad, black, triangular; proboscis 1.40 mm long, cylindrical, brown scally; labium 1.15 mm long, slender; labellum 0.25 mm long, brown; palpi 0.35 mm long, three segmented, shorter than proboscis; palpifer brown, palpus brown, scally; mandibles and maxillae as long as proboscis, yellow, straight, palpi with scaling, proboscis thick and about the length of fore femur. Proboscis entirely dark and without white or pale ring.

Antenna (Figure 105)

1.65 mm long, blackish brown, 13 segmented, pilose, hairy; scape 0.05 mm long, 0.15 mm broad, brown; pedicel 0.25 mm long, 0.10 mm broad, rounded, brown, flagellum 1.35 mm long, 15 segmented.

Flagellar Formula

3 L/W = 4, 9 L/W = 4.75, L 3/9 = 1.05, W 3/9 = 1.25, A = 2.76.

Thorax

1.10 mm long, 1.30 mm broad, golden brown, laterally compressed; scutum brown, posterior pronotal lobe brown; scutellum (Figure 106) 0.5 mm long (Figure 5), 0.6 mm broad, brown, rounded, with silvery white scales; sternopleuron brown; mesepimeron broad, flat, with white patches; mesothoracic spiracles rounded; pronotum lack of satae. Half circular white mark and some white spots are present on thorax, blackish and white spots are absent on mesonotum.

Forewing (Figure 107)

2.70 mm long, 0.65 mm broad, dark wings not spotted, wing membrane not clouded, unspotted, all scales dark;

subcosta straight, 2.20 mm long, reaching the costa; media straight, bifurcated, slightly curved at end; cubitus bifurcated.

Halter (Figure 108)

0.50 mm long, 0.09 mm broad, yellow, rounded in lateral view.

Hind Leg (Figure 109)

7.90 mm long, slender, longer than body; coxa 0.20 mm long, yellowish, broad, with 4 large and 7 small coxal bristles; trochanter 0.10 mm long and 0.10 mm broad, rounded, hard, brown; femur 1.65 mm long, cylindrical; tibia 2.15 mm dark brown, with white ring, mid tibia pale at posterior side; tarsus 3.80 mm long, five segmented; 1st tarsal segment 1.65 mm long, 2nd tarsal segment 1.15 mm long, 3rd tarsal segment 0.60 mm long, 4th tarsal segment 0.30 mm long, 5th tarsal segment 0.10 mm long, with dark and white bands, 1st hind tarsal segment less than tibia; claw toothed, five white rings on hind tarsi at basal side, two white rings are present on foreleg and middle, hind tarsi with white ring on 4th segment very wide covering nearly the whole segment, mid femura with preapical white spots, venter of abdomen without orange marking.

Other Legs

Special marks : similar.

Abdomen (Figure 110)

2.25 mm long, 0.65 mm broad, blackish with snow white marking and basal bands, slender, pointed, black and white transverse bands, white patches of scales, laterotergites, sternal plate dirty white, segment VIII broader and not completely retractile, sternite VIII large, prominent tergal plates dark blackish brown; anal cerci short and broad, 0.03 mm long and 0.04 mm broad, brown, flat, hairy.

Colour

Black : Eyes, head.

Brown : Proboscis, scape, thorax.

Blackish Brown : Abdomen.

Yellow : Mandibles, maxillae, wing.

Male : 3.10 mm long, slender, smaller than female; antenna plumose brushy, phytophagous.

Host : Human and Cattle.

Host Plant : Unknown.

Holotype : Female, India, Maharashtra, Kagal coll. Jagtap, M.B., 12-VII-2008 head, antenna, hind leg, wing and abdomen mounted and pinned, labeled as above.

Paratype : 3 ♂, 14 ♀, sex ratio (M:F) 1:4.66 coll. Jagtap, M.B. Jun. 2006 to Dec. 2009.

Etymology : The species name *Aedes* (Finalaya) *rajashri* sp. nov. is given because species is found in Kagal *i.e.* birth place of His Highness Chattrapati Rajashri Shahu Maharaj

Distributional Record

2 ♂ 4 ♀ Kolhapur 24-VI-2006; 1♂, 3 ♀, Jaysingpur, 12-XI-2006; 0 ♂ 4 ♀ Kagal, 12-VII-2008; 0 ♂, 3 ♀, Jaysingpur 12-XII-2009.

Remarks

According to the key of Barraud (1934) taxonomically this species runs close to *Aedes* (Finalaya) *subsimilis* by having following characters.

1. Tarsi with white markings.
2. Abdominal venter not orange.
3. Hind tarsi with one white basal ring on segments

However, it differs from the above species by having following characters.

1. Five white rings on hind tarsi.
2. Two white rings are present on foreleg and mid leg.
3. Abdominal venter dark brown in colour.
4. Mesonotum blackish with white spots absent.
5. Mid tibia pale at posterior side
6. Scutellum shape (rounded) and size (0.5 mm long and 0.6 mm broad).
7. Halter shape (rounded) and size (0.50 mm long and 0.09 mm broad).
8. Flagellar formula :
   3 L/W = 4, 9 L/W = 4.75, L 3/9 = 1.05, W 3/9 = 1.25, A = 2.76.
9. Phyllogenetically it runs close to *Ochlerotatus togoi*. However, it differs from 16 species by having neighbourhood joining branch length = 7.19385402.

Genus *Culex* Linnaeus 1758

Linnaeus 1758 : 602. Michener 1944; 263 (*Cibarial armature*). Senevet 1947 : 212 Larvalkey, Africa), Senevet and Andarelli 1954a) : 36 (Key, N. Africa). Sicart 1954 : 27 (Key to pupae, Tunisia). Bram 1967b : 1 (Thailand). Bram 1969 : 9 (Tax., S.E. Asia).

The genus *Culex* is raised by Linnaeus in 1758. From various part of the world more than 400 species of this genus have been reported (Stone, *et al.* 1959). The genus is characterized by:

1. In female buccopharyngeal armature present.
2. Palpi shorter than proboscis.
3. Pulvilli well developed.
4. No spiracular or post spiracle bristles.
5. Well developed mesonotal bristles.
6. Scutellar and mesonotal scales narrow.
7. Tarsi without pale ring.
8. Scutellum trilobed with three tuft of hairs on the lobe.
9. Pronotum without setae.
10. Wings unspotted with only dark scales.
11. Simple and curved claws.
12. Palpi four segmented, may be entirely dark or ringed with white.
13. Abdomen slightly tapering blunt and covered with broad scales.
14. Very small anal cerci.

Barraud (1934) divided the genus *Culex* into seven subgenera given below:

1. *Culex* Linnaeus 1758.
2. *Lutzia* Theobald, 1903,

3. *Neoculex* Dyar, 1905,
4. *Lophoceratomyia* Theobald, 1905,
5. *Culiciomyia* Theobald, 1907,
6. *Barraudius* Edwards, 1921 and
7. *Mochthogenes* Edwards 1930.

58 species have been reported under above subgenera from India. The sub genus *Culex* contain 23 species, *Lutzia* contain 4 species, *Barraudius* 2, *Neoculex* 2, *Mochthgenes* 5, *Lophoceratomyia* 10 and *Culiciomyia* 7 species.

Key to the Species of Subgenera *Culex*

1. Proboscis with pale ring in middle .... *Culex*

   Proboscis without a pale ring 2

2. First hind tarsal segment distinctly shorter
   than tibia .... *Barraudis*

   First hind tarsal segment about as long as
   tibia or longer 3

3. Dorsal surface of head mainly covered with
   flat scales .... *Mochthogenes*

   Dorsal surface of head mainly covered
   with narrow scales .... *Neoculex*

Subgenus *Culex* Linnaeus

Linnaeus 1758 : 602. Logotype : *Pipiens* Linnaeus (Latrille 1810 : 442). *Heteronycha* Lynch Arribalzaya 1891a : 373, 1891b : 155 Haplotype : *dolosa* Lynch Arribalzaya. *Lasioconops* Theobald 1903a : 8 Haplotype : poicilipes Theobald. *Heptaphlebomyia* Theobald 1903b : 336 Haplotype : *Simplex* Theobald. Culex, subgenus *Transculicia* Dyar : 1917 : 184. Haplotype : eleuthera Dyar. Bram 1967 a: 1 (New world). Belkin, Sebick, and Heinemann 1969 : 9 (type-Loc-info, S. America). 1971 : 1 (type-Loc-info, Brazil). Chen 1972 : 282 (Cibarium key to some *Culex* spp.) Cupp and Ibrahim 1973 : 277 (*Pipens* complex).

Syst. Nat. ed. x, p. 602, Genotype, *C. pipiens* Linn. *Leucomyia* Theobald, 1907, M.C. iv, p. 372. Genotype *C. gelidus* Theo.

Oculeomyia Theobald, 1907, M.C. iv, p. 515.

Genotype *C. bitaeniorhynchus* Giles (as *O. sarawati* Theo.)

Theobaldiomyia Brunetti, 1912 Rec. Ind. Mus. iv, p. 462 (nom. for *Leucomyia* Theo.) (Edwards, 1932, p. 200).

The subgenus *Culex* shows following features:

1. Body size moderate.
2. Scutellar scales narrow.
3. Head with narrow scales on vertex, flat scales at sides.
4. Last two segments of palpi upcurved and hairy.
5. Palpi of male always longer than proboscis.
6. Male antenna without scale tuft.
7. Hypopysium (male) without scales.
8. Stylet without spiny crest.
9. Paraprocts with dense tuft of spines or hairs at crown.
10. Phallosome divided into lateral portions with various teeth or lobes.
11. In female buccopharyngeal armature with one row of teeth, teeth may be sharp or blunt.

Key to the Species of Subgenera *Culex*

1. Proboscis and tarsi with pale rings
   (tarsal rings sometimes faint) : no
   lower mesepimeral bristle 2

   Proboscis without pale ring (sometimes
   indistinctly ringed in ♂) : tarsi entirely
   dark; lower mesepimeral bristle present 13

2. Wings with costa dark unless at tip 3

   Wings with three pale spots on costa
   (including one at tip) 12

3. A yellowish area at tip of wing; body
   and legs largely yellow .... *epidesmus*,

   Tarsi dark with 2 pale rings at 4,5th
   segment, 2 golden stripes on thorax .... *malhari*

   No yellowish area at wing-tip; body and
   legs not usually extensively yellow 4

4. Wings speckled with pale scales,
   which are usually numerous .... *bitaniorhynchus*

   Wing-scales all dark 5

5. Scales on anterior 2/3 of mesonotum
   mainly or all white or pale ochreous 6

   Mesonotal scales mainly dark or with
   indefinite pale mottling 8

6. Abdominal tergites spically banded;
   mesonotal scales pale ochreous .... *sinensis*

   Abdominal tergites basally banded;
   mesonotal scales white 7

7. Wings with broad scales on veins 1, 3,
   and 3; mid and hind tibiae with
   pale lines .... *uchitmorei*

   Wings without unusually broad scales;
   tibiae not lined .... *gelidus*

8. Abdominal tergites with distinct
   apical pale markings .... *cornutus*

   Markings of tergites mainly or entirely basal 9

9. Middle femur with a pale stripe in front .... *edwardsi*

   Middle femur not striped 10

10. Femora speckled with pale scales, especially
    anterior surface of mid-femur .... *sitiens,*

    Femora without any sprinkling of pale scales 11

11. Mesonotum with light and dark scales
    mixed in varying proportions, sometimes
    forming an indefinite .... *(barraudi,*

    pattern, but at least with light scales .... *whitei,*

    round front margin .... *vishnui,*

    Mesonotal scales uniformly
    dark brown .... *tritaeniorhynchus,*

12. Pale spot at middle of wing involving
    only costa and subcosta .... *mimeticus*

    Pale spot at middle of wing usually
    extending over vein 1 .... *mimulus*

13. Pleurae devoid of scales; proboscis all
    black : ♂ palpi entirely dark scaled .... *nilgiricus*

Pleurae with patches of broad scales;
proboscis pale beneath in middle;
last two segments of ♂ palpi with
a white line beneath, or at least with
a white spot at the base of each 14

14. Fore and mid-femora and all tibiae
conspicuosly striped in front 15

 Fore and mid-femora all dark in front 16

15. Hind femur with a brown line beneath
on distal 1/3 .... *theileri*

 Hind femur pale beneath from base
to knee .... *vagans*

16. Integument of pleurae uniformly coloured;
abdominal tergites basally banded 17

 Integument of pleurae with bare blackish-
brown areas situated immediately above
and below a conspicuous patch of white
scales in middle 19

17. Two terminal segments, or at least penultimate
segments, of ♂ palpi with white line beneath 18

 Terminal segments of ♂ palpi with white
basal spots beneath; very small species .... *hutchinsoni*

18. Mid and hind tibiae with more or less
obvious pale stripe on outer side; abdominal
bands white : a patch of white scales
behind prothoracic spiracle .... *univittatus,*

 Mid- and hind tibiae dark, except small
pale spot at tip : abdominal bands
creamy : no post-spiracular scales .... *fatigans*

Proboscis and legs unbanded, No
ornamentation, dark tarsi and Rounded
abdominal bands ....*quinquifasciatus*

Thorax is light redish colour Palpi is
not 1/6ᵗʰ of proboscis .... *malkapuri*

19. Abdomen unbanded ....*fuscocephalus*

Abdomen banded ....*fuscitarsis*

Culex (Culex) *quinquifasciatus* Say, 1823

Female (Figure 150)

4.20 mm long, 0.75 mm broad, small sized, dark brown, without ornamentation; head 0.65 mm long, 0.45 mm broad, blackish brown; antenna 1.55 mm long, brown; thorax 1.50 mm long, 0.75 mm broad, brown; wing 3.75 mm long, 0.65 mm broad, yellow; hind leg 6.50 mm long, brown; abdomen 2.10 mm long, 0.60 mm broad, blackish brown. There is no ornamentation on any part of its body.

Head (Figure 115)

0.65 mm long, 0.45 mm broad, triangular, blackish brown, dorsal surface of head with flat scales; compound eyes black, round; ocular space 0.20 mm, interocular distance 0.15 mm; vertex and nape covered with narrow golden brown scales, brown, clypeus triangular, brown; nape 0.10 mm long, tubular brown, short; proboscis 1.80 mm long and unbanded, cylindrical, curved, brown, scally; labium 1.55 mm long, slender, curved, scally; labellum 0.25 mm long, densely scally, forked, pointed; palpi 0.20 mm long and brown about 1/6ᵗʰ length of proboscis, very short, four segmented, hairy, brown, shorter than proboscis.

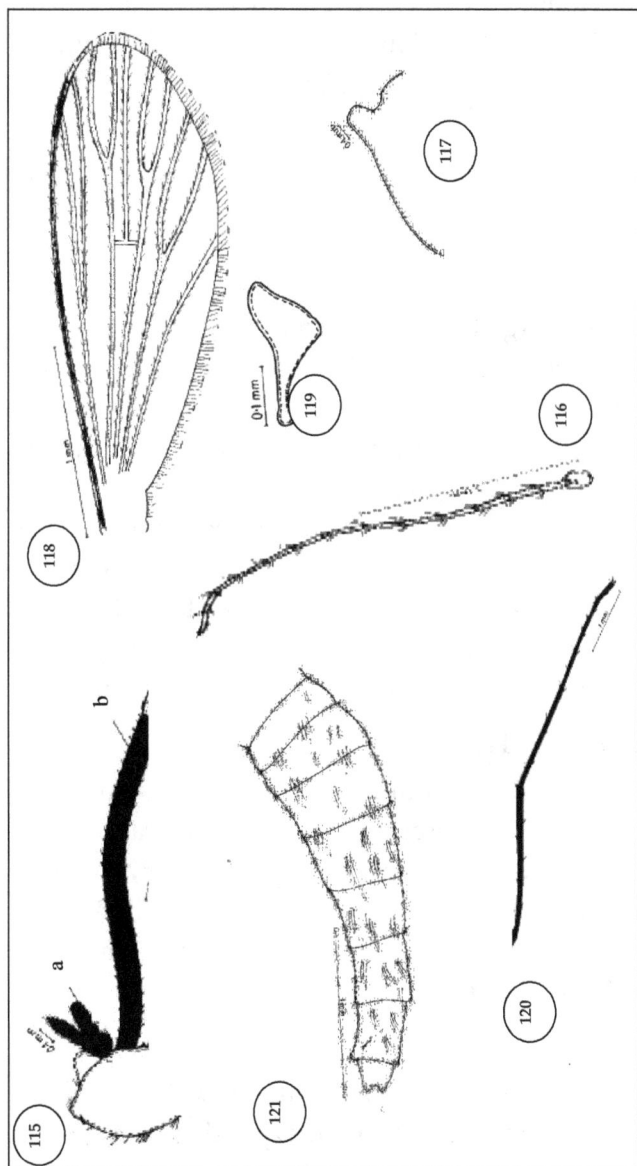

Plate 19—*Culex* (Culex) *quinquifasciatus* sp. nov.

Figure 115: Head-Palpi (a) and Proboscis (b); **Figure 116**: Antenna; **Figure 117**: Scutellum; **Figure 118**: Forewing; **Figure 119**: Halter; **Figure 120**: Hind leg; **Figure 121**: Abdomen

Antenna (Figure 116)

1.55 mm long, 15 segmented, brown, hairy, pilose; dark brown; pedicel 0.10 mm long, 0.09 mm broad, rounded, brown; flagellum 1.45 mm long, 15 segmented.

Flagellar Formula

3 L/W = 2.66, 9 L/W = 3.00, L 3/9 = 1.06, W 3/9 = 1.2, A = 1.98.

Thorax

1.50 mm long, 0.75 mm broad, brown, laterally compressed, dorsal brown stripes present; plurae uniformly not brown, mesonotum scutellum covered with narrow golden brown scales. From front of mesonotum two submedian dark bare lines run back a short distances. Scutum half moon shaped, brown smooth; scutellum (Figure 117) 0.35 mm long, 0.20 mm broad, triangular in lateral view, blackish brown; postnotum flat, not hairy; sternopleuron triangular, brown; metathoracic spiracle, rounded, yellow; lower mesepimeron bristles absent.

Forewing (Figure 118)

3.75 mm long, 0.65 mm broad, unspotted, with dark scales, scales 0.03 mm long, veins sparsely scally; costa straight, dark blackish brown; subcosta straight, 2.70 mm long, reaching the costa; radius straight slightly curved, simple, without cross veins; cubitus bifurcated; anal vein curved and short, extend to wing margin.

Halter (Figure 119)

0.25 mm long, 0.20 mm broad, without scale, yellowish, triangular in lateral view, brown, expanded at tip; stalk faint yellow.

Hind Leg (Figure 120)

6.35 mm long, brown or nearly black, elongated longer than body; coxa 0.15 mm long; trochanter 0.10 mm long, yellowish, rounded; femur 1.40 mm long with very small knee spots, cylindrical, yellowish, scally; tibia 2.05 mm long usually marked with narrow yellowish ring, slender, yellow, with two spurs, spurs equal; tarsus 2.65 mm long and dark, five segmented, 1st tarsal segment, 1.20 mm long, 2nd tarsal segment 0.95 mm long, 3rd tarsal segment 0.25 mm long, 4th tarsal segment 0.15 mm long, 5th tarsal segment 0.10 mm long, claw simple curve; empodium and pulvillus, small; legs dark brown and unbanded.

Other Legs

Special marks : similar.

Abdomen (Figure 121)

2.10 mm long, 0.60 mm broad, dorsum of abdomen with pale bands, tergite blackish brown, tergite I almost entirely covered with long yellow hairs, sternum yellowish brown; post genital plate 0.15 mm long, 0.09 mm broad, brown, hairy, anal cerci 0.09 mm long, 0.05 mm broad, hairy, brown, integument of plurae with dark spots.

Colour

Black	: Eyes.
Brown	: Vertex, clypeus, proboscis, antenna, thorax, legs.
Dark brown	: Head, abdomen.
Yellow	: Mandibles, maxillae, wing.
Yellowish brown	: Halter.

Male	: 2.75 mm long, smaller than female; antenna Plumose, brushy, phytophagous.
Host	: Cattle.
Host Plant	: Unknown.
Holotype	: Female, India, Maharashtra, Koregaon coll. Jagtap, M. B., 3-V-2007; head, antenna, hind leg, abdomen mounted on slide, labeled as above.
Paratype	: 119♂, 381 ♀, sex ratio (M:F) 1:3.20 coll. Jagtap, M. B., Jan. 2006 to Dec. 2009.

Distributional Record

4 ♂ 12♀ Jath, 5-III-2006; 3 ♂ 10 ♀ Saswad, 12-III-2006; 2 ♂, 8 ♀, Tasgaon, 9-IV-2006; 2♂ 8 ♀ Kolhapur, 24-VI-2006; 3 ♂ 10 ♀ Jaysingpur, 9-VII-2006; 5 ♂ 15 ♀ Miraj, 11-VI-2006; 4 ♂ 15 ♀ Kolhapur 24-VI-2006, 6 ♂ 14 ♀ Kagal 12-VIII-2006; 6 ♂ 17 ♀ Malakapur, 11-II-2007; 3 ♂ 17♀ Koregaon, 3-V-2007; 4 ♂ 14 ♀ Shirala, 14-VI-2007; 8 ♂, 17 ♀, Pune, 25-VII-2007; 5 ♂ 16 ♀ Satara, 23-XI-2007; 4 ♂, 9 ♀, Vita, 23-III-2008; 2 ♂ 9 ♀ Ajara, 26-IV-2008; 7 ♂ 25 ♀ Kagal, 12-VII-2008; 4 ♂ 15♀ Medha, 20-VII-2008; 3 ♂, 13 ♀, Bhor, 10-VIII-2008; 5 ♂ 16 ♀ Mhaswad, 23-VIII-2008; 7 ♂ 19 ♀ Saswad, 11-X-2008; 4 ♂ 14 ♀ Junner, 22-II-2009; 4 ♂ 10 ♀ Vita, 25-IV-2009; 2 ♂, 15♀, Kolhapur, 14-VI-2009; 3 ♂ 15 ♀ Patan, 11-VII-2009; 5 ♂ 14 ♀ Wai, 9-VIII-200; 6 ♂ 15 ♀ Bhor 29-XI-2009; 8 ♂, 19♀, Jaysingpur 12-XII-2009.

Remark

According to the key of Barraud (1934) this species runs close to *Culex quinquefasciatus* by having following characters.

1. The proboscis and legs are unbanded. There is no ornamentation on any part of its body.
2. Dark tarsi and rounded abdominal bands.
3. From front of mesonotum two submedian dark bare lines run back a short distances.

However, some following additional characters have been observed:

1. Scutellum shape (triangular) and size (0.35 mm long and 0.20 mm broad).
2. Flagellar formula:

   3 L/W = 2.66, 9 L/W = 3.00, L 3/9 = 1.06, W 3/9 = 1.2, A = 1.98.
3. Phyllogenetically it runs close to *Culex quinquifaciatus*. However, it differs from 17 species by having branch length = 0.01732766.

Culex **(Culex)** *malhari* **sp.nov.**

Female (Figure 151)

4.55 mm long, 0.70 mm broad, small sized, yellowish coloured mosquito, without ornamentation; head 0.75 mm long, 0.50 mm broad, blackish brown; antenna 1.60 mm long, brown; thorax 1.60 mm long, 0.70 mm broad, brown; wing 3.90 mm long, 0.75 mm broad, yellow; hind leg 6.70 mm long, brown; abdomen 2.20 mm long, 0.65 mm broad, blackish brown. There is no ornamentation on any part of its body.

Head (Figure 122)

0.75 mm long, 0.50 mm broad, triangular, yellowish brown, dorsal surface of head with narrow yellowish and brown scales; compound eyes black, round; ocular space 0.20 mm, interocular distance 0.15 mm; vertex and nape covered with narrow golden brown scales, brown, clypeus triangular, brown; nape 0.09 mm long, tubular brown, short; proboscis 1.70 mm long with yellow ring at middle, cylindrical, curved, brown, scaly; labium 1.50 mm long, slender, curved, scally; labellum 0.25 mm long, densely scally, forked, pointed; palpi 0.20 mm long and brown about 1/6th length of proboscis, very short, four segmented, hairy, brown, shorter than proboscis; mandible maxillae slender, longer, straight, yellow.

Antenna (Figure 123)

1.60 mm long, 15 segmented, brown, hairy, pilose; dark brown; pedicel 0.10 mm long, 0.09 mm broad, rounded, brown; flagellum 1.50 mm long, 15 segmented.

Flagellar Formula

3 L/W = 2.66, 9 L/W = 3.00, L 3/9 = 1.06, W 3/9 = 1.2, A = 1.98.

Thorax

1.60 mm long, 0.70 mm broad, yellow, laterally compressed, dorsal brown stripes present; plurae uniformly not brown, mesonotum scutellum covered with narrow golden brown scales. From front of mesonotum two submedian dark bare lines run back a short distances. Scutum half moon shaped, brown smooth; scutellum (Figure 124) 0.35 mm long, 0.25 mm broad, scutellar scales not narrow, triangular in lateral view, blackish brown;

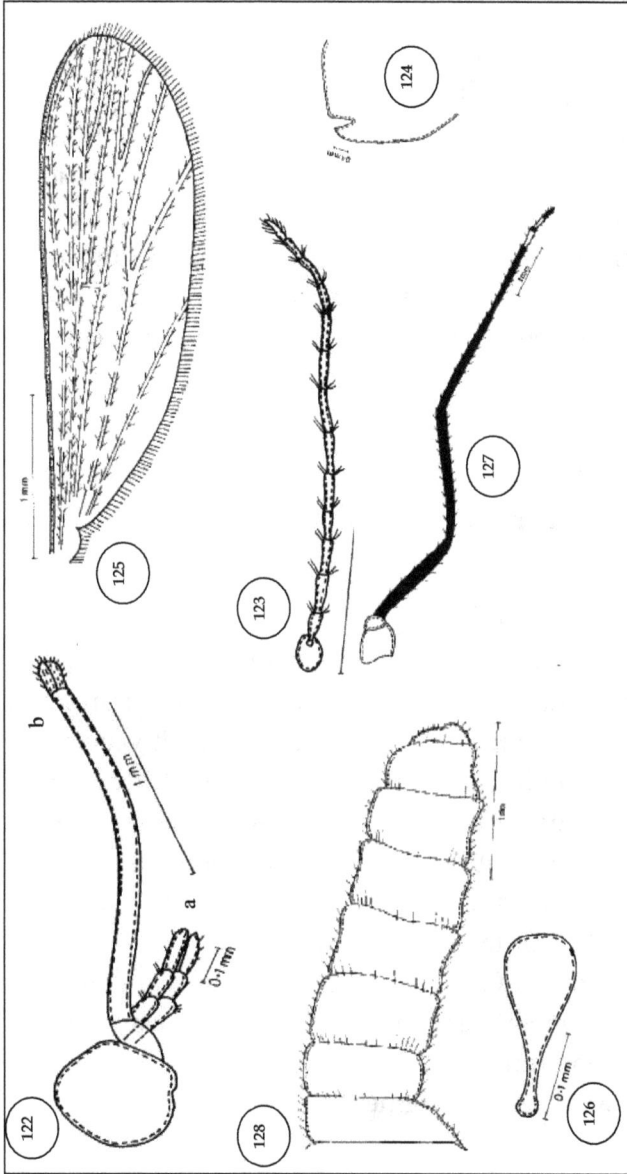

Plate 20—*Culex* (Culex) *malhari* sp. nov.

Figure 122: Head-Palpi (a) and Proboscis (b); **Figure 123:** Antenna; **Figure 124:** Scutellum; **Figure 125:** Forewing; **Figure 126:** Halter; **Figure 127:** Hind leg; **Figure 128:** Abdomen

postnotum flat, not hairy; sternopleuron triangular, brown; metathoracic spiracle, rounded, yellow; lower mesepimeron bristles absent.

Forewing (Figure 125)

3.90 mm long, 0.75 mm broad, unspotted, with yellowish colour, scattered yellow scales 0.03 mm long along costa and vein 1, veins sparsely scally; costa straight, dark blackish brown; subcosta straight, 2.75 mm long, reaching the costa; radius straight slightly curved, simple, without cross veins; cubitus bifurcated; anal vein curved, extend to wing margin, anal vein short.

Halter (Figure 126)

0.30 mm long, 0.25 mm broad, without scale, yellowish, triangular in lateral view, brown, expanded at tip; stalk faint yellow.

Hind Leg (Figure 127)

6.70 mm long, yellowish in colour, elongated longer than body; coxa 0.15 mm long; trochanter 0.10 mm long, yellowish, rounded; femur 1.55 mm long with very small knee spots, cylindrical, yellowish, scally; tibia 2.20 mm long usually marked with narrow yellowish ring, slender, yellow, femora and tibia yellowish with numerous scattered brown scales; tarsus 2.70 mm long, five segmented, 1st tarsal segment, 1.20 mm long, 2nd tarsal segment 1.00 mm long, 3rd tarsal segment 0.25 mm long, 4th tarsal segment 0.15 mm long, 5th tarsal segment 0.10 mm long, claw simple curved; empodium and pulvillus, small; Tarsi dark with 2 pale rings at 4th and 5th segment.

Other Legs

Special marks : similar.

Abdomen (Figure 128)

2.20 mm long, 0.65 mm broad, dorsum of abdomen with pale bands, tergite golden brown, tergite I almost entirely covered with long yellow hairs, sternum yellowish brown; post genital plate 0.15 mm long, 0.08 mm broad, brown, hairy, anal cerci 0.08 mm long, 0.05 mm broad, hairy, brown, integument of plurae with dark spots.

Colour

Black	: Eyes.
Brown	: Proboscis.
Dark brown	: Antenna.
Yellow	: Thorax, legs, mandibles, maxillae, wing.
Yellowish brown	: Halter, Head, Abdomen, Vertex, clypeus.
Male	: 2.80 mm long, smaller than female; antenna Plumose, brushy, phytophagous.
Host	: Cattle.
Host Plant	: Unknown.
Holotype	: Female, India, Maharashtra, Saswad coll. Jagtap, M. B., 11-X-2008; head, antenna, hind leg abdomen mounted on slide, labeled as above.
Paratype	: 36 ♂, 80 ♀, sex ratio (M:F) 1:2.22 coll. Jagtap, M. B., Jan. 2006 to Feb. 2009.

Etymology : The species name is *Culex* (Culex) *Malhari* sp. nov. refers to the God Malhari which is situated near the collection site of mosquito *i.e.* Saswad district Pune, Maharashtra, India.

Distributional Record

2♂, 4♀, Koregaon, 3-V-2007; 5♂ 7 ♀ Shirala, 14-VI-2007; 1 ♂ 5 ♀ Pune, 25-VII-2007; 4♂, 7 ♀ Baramati, 13-I-2008; 2 ♂ 5 ♀ Ajara, 26-IV-2008; 2 ♂, 10 ♀, Vita, 23-III-2008; 3 ♂, 9♀, Vita, 23-III-2008; 3♂ 9 ♀ Saswad, 11-X-2008; 2♂, 4♀, Jaysingpur, 21-IX-2008; 3 ♂ 7 ♀ Saswad, 11-X-2008; 5 ♂ 8 ♀ Koregaon, 10-I-2009; 4♂ 9 ♀ Bhor, 21-II-2009.

Remark

According to the key of Barraud (1934) this species runs close to *Culex epidesmus* Theobald 1910 by having following characters.

1. Proboscis and tarsi with pale rings.
2. No lower mesepimeral bristles.
3. Yellowish area at tip of wing, body and legs largely yellow.

However, it differs from the following characters.

1. Scutellar scales not narrow.
2. Tarsi dark with 2 pale rings at 4th and 5th segment.
3. Two golden stripes on thorax.
4. Scutellum shape (triangular) and size (0.35 mm long and 0.25 mm broad).

5. Flagellar formula:

    3 L/W = 2.66, 9 L/W = 3.00, L 3/9 = 1.06, W 3/9 = 1.2, A = 1.98.

Culex (Culex) *malkapuri* sp.nov.

Female (Figure 152)

    4.00 mm long, 0.70 mm broad, small sized, dark brown, without ornamentation; head 0.60 mm long, 0.45 mm broad, blackish brown; antenna 1.55 mm long, brown; thorax 1.45 mm long, 0.70 mm broad, brown; wing 2.80 mm long, 0.65 mm broad, yellow; hind leg 6.90 mm long, brown; abdomen 1.90 mm long, 0.70 mm broad, blackish brown.

Head (Figure 129)

    0.60 mm long, 0.45 mm broad, triangular, dark brown narrow scales on head, some broader scales are present on head; compound eyes black, round; ocular space 0.22 mm, interocular distance 0.11 mm; vertex flat, brown, clypeus triangular, brown; nape 0.10 mm long, tubular brown, short; proboscis 1.75 mm long brown, without pale ring at middle part, cylindrical, curved, brown, scaly; labium 1.45 mm long, slender, curved, scally; labellum 0.30 mm long, densely scally, forked, pointed; palpi 0.25 mm long and not 1/6th of proboscis, very short, four segmented, hairy, brown, shorter than proboscis; mandible maxillae slender, longer, straight, yellow.

Antenna (Figure 130)

    1.55 mm long, 15 segmented, dark brown, hairy, tuff of scales are more on antenna, pilose; dark brown; pedicel 0.12 mm long, 0.09 mm broad, rounded, brown; flagellum 1.50 mm long, 15 segmented.

Flagellar Formula

3 L/W = 2.8, 9 L/W = 2.16, L 3/9 = 1.07, W 3/9 = 0.83, A = 1.71.

Thorax

1.45 mm long, 0.70 mm broad, light reddish colour, with golden brown narrow curved scales, laterally compressed, 2 dorsal brown lines present; plurae uniformly not brown; scutum half moon shaped, basal abdominal bands curved, brown smooth; scutellum (Figure 131) 0.35 mm long, 0.23 mm broad, triangular in lateral view, blackish brown; postnotum flat, not hairy; sternopleuron triangular, brown; metathoracic spiracle, rounded, yellow; lower mesepimeron bristles absent.

Forewing (Figure 132)

2.80 mm long, 0.65 mm broad, unspotted, with dark scales, scales 0.03 mm long, veins sparsely scally; wings with submarginal cell longer and narrower than 2nd posterior cell, scales on 6th vein scanty, costa straight, dark blackish brown; subcosta straight, 2.73 mm long, reaching the costa; radius straight slightly curved, simple, without cross veins; cubitus bifurcated; anal vein curved, extend to wing margin, anal vein short.

Halter (Figure 133)

0.20 mm long, 0.15 mm broad, without scale, yellowish, triangular in lateral view, brown, expanded at tip; stalk faint yellow.

Hind Leg

6.90 mm long, legs are brown, elongated longer than body; coxa 0.20 mm long; trochanter 0.15 mm long, yellowish, rounded; femur 1.70 mm long, cylindrical, pale

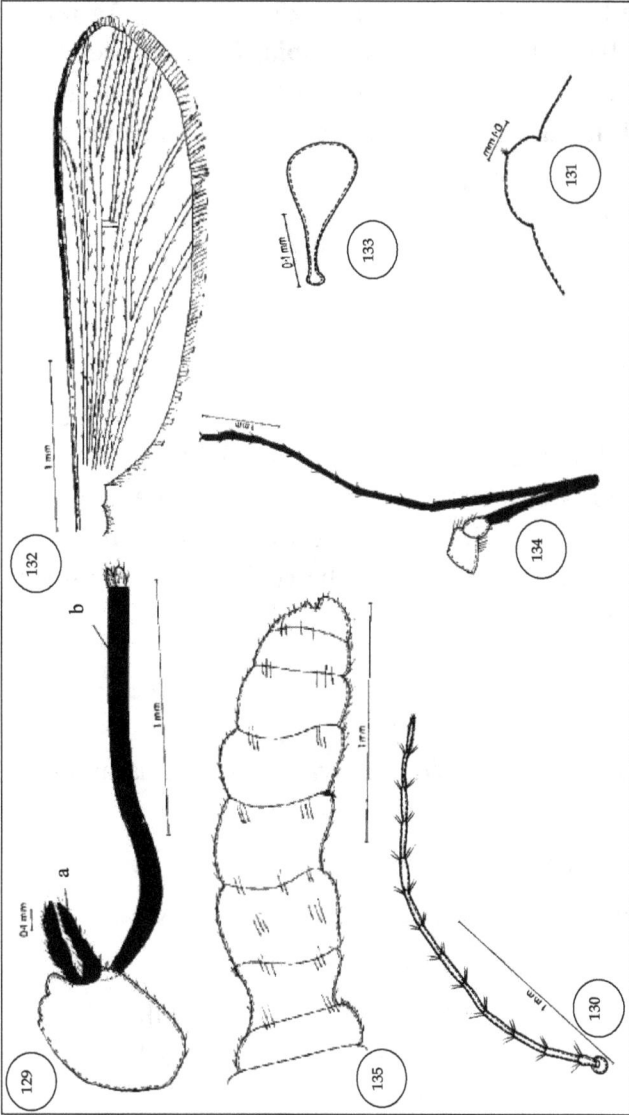

Plate 21–*Culex* (Culex) *malkapuri* sp. nov.

Figure 129: Head-Palpi (a) and Proboscis (b); **Figure 130**: Antenna; **Figure 131**: Scutellum; **Figure 132**: Forewing; **Figure 133**: Halter; **Figure 134**: Hind leg; **Figure 135**: Abdomen

or yellowish, scally; tibia 2.20 mm long, slender, yellow, with two spurs, spurs equal; tarsus 2.65 mm long, five segmented, 1st tarsal segment, 1.25 mm long, 1st tarsal segment distinctly shorter than tibia, 2nd tarsal segment 0.90 mm long, 3rd tarsal segment 0.25 mm long, 4th tarsal segment 0.15 mm long, 5th tarsal segment 0.10 mm long, claw simple curve; empodium and pulvillus, small; legs not dark brown.

Other Legs

Special marks : similar.

Abdomen (Figure 134)

1.95 mm long, 0.70 mm broad, dark brown with basal white or pale curved bands, tergite blackish brown, sternum yellowish brown; lateral patches absent, post genital plate 0.13 mm long, 0.09 mm broad, brown, hairy, anal cerci 0.09 mm long, 0.05 mm broad, hairy, brown, integument of plurae with dark spots.

Colour

Black	: Eyes.
Brown	: Vertex, clypeus, proboscis, antenna, thorax, legs.
Dark brown	: Head, abdomen.
Yellow	: Mandibles, maxillae, wing.
Yellowish brown	: Halter.
Male	: 2.85 mm long, smaller than female; antenna plumose, brushy, phytophagous.
Host	: Cattle.

Host Plant	: Unknown.
Holotype	: Female, India, Maharashtra, Malkapur coll. Jagtap, M. B., 12-VII-2007; head, antenna, hind leg abdomen mounted on slide, labeled as above.
Paratype	: 23♂, 60 ♀, sex ratio (M:F) 1:2.60 coll. Jagtap, M. B., April 2006 to Dec. 2009.
Etymology	: The species name *Culex* (Culex) *malkapuri* sp. nov. refers to the collection site of mosquito *i.e.* Malkapur, district Kolhapur, Maharashtra, India.

Distributional Record

2♂, 6♀, Tasgaon, 9-IV-2006; 2♂, 6♀, Kagal, 12-VIII-2006, 2♂ 5 ♀ Malakapur, 12-VII-2007; 5♂, 11 ♀, Kolhapur, 8-VIII-2007, 2 ♂ 7 ♀ Pune, 20-XII-2007; 5 ♂, 12 ♀, Vita, 23-III-2008; 5 ♂, 9 ♀, Bhor, 10-VIII-2008; 0 ♂, 4♀, Jaysingpur 12-XII-2009.

Remark

According to the key of Barraud (1934) this species runs close to *Culex* (Culex) *fatigans* Wiedemann 1828 by having following characters.

1. The proboscis and legs are unbanded.

2. No ornamentation of body.

3. Dark tarsi and rounded abdominal bands.

However, it differs from above species by having following characters:

1. Thorax is light reddish colour.
2. The dark bands of tergites, labium, femora and tibia are covered with black scales.
3. Palpi is not 1/6th of proboscis.
4. On sternite lateral patches absent.
5. Scutellum shape (triangular) and size (0.35 mm long, 0.23 mm broad).
6. Flagellar formula:
   3 L/W = 2.8, 9 L/W = 2.16, L 3/9 = 1.07, W 3/9 = 0.83, A = 1.71.
7. Phyllogenetically it runs close to *Culex pipiens complex*. However, it differs from 16 species by having NHJ branch length = 0.58109042.

Sub Genus: *Barraudius* Edwards 1921

Edwards 1921 a described the subgenus *Barraudius*. He reported two species from India. (Barraud, 1934; Sathe and Girhe, 2002). Recently, Sathe and Tingare, 2010 added two species under this subgenus. This subgenus is characterized by having following features:

1. Scales on vertex of head 1, Apn and scutellum all narrow.
2. Small mosquitoes without ornamentation.
3. Pulvilli well developed.
4. Segment one of hind tarsi distinctly shorter than tibia.
5. Paraproct with hairs on spines at crown, no lateral arms, *Bull. Ent. Res.*, xii, p. 322. Genotype C. *pusillus* Maeq.

Key to the Species of Subgenus *Barraudius* Edwards

1. Palpi about 1/6 length of proboscis; first hind tarsal segment equal to tibia .... *modestus*

   Palpi small, not 1/6 length of proboscis; first hind tarsal segment shorter than tibia, hind femur slightly longer than tibia .... *krishnai* sp.nov.

   Proboscis with apical large dark Band, coxal bristles 3 large and 5 small .... *kalambae* sp.nov.

   Legs are not dark brown, Proboscis without apical dark band, Hind femur shorter than tibia .... *mirjensis* sp. nov.

   3 black strips on thorax, wings not dark, Abdomen without pale boarder on either side Antenna without tuft of scales .... *satarensis* sp. nov.

Culex (Barraudius) *mirjensis* sp. nov.

Female (Figure 153)

   3.80 mm long, 0.70 mm broad, small sized, dark brown, without ornamentation; head 0.50 mm long, 0.40 mm broad, blackish brown; antenna 1.60 mm long, brown; thorax 1.40 mm long, 0.70 mm broad, brown; wing 2.90 mm long, 0.60 mm broad, yellow; hind leg 6.60 mm long, brown; abdomen 1.90 mm long, 0.70 mm broad, blackish brown.

Head (Figure 136)

   0.50 mm long, 0.45 mm broad, triangular, brown narrow scales on head, some broader scales are present on head, compound eyes black, round; ocular space 0.20 mm, interocular distance 0.10 mm; vertex flat, brown, clypeus triangular, brown; nape 0.10 mm long, tubular brown, short; proboscis 1.70 mm long, without pale ring at middle part and apical dark band, cylindrical, curved, brown, scaly; labium 1.45 mm long, slender, curved, scally; labellum 0.25 mm long, densely scally, forked, pointed; palpi 0.20mm long, very short not 1/6th of proboscis, four segmented, hairy, brown, shorter than proboscis; mandible maxillae slender, longer, straight, yellow.

Antenna (Figure 137)

1.60 mm long, 15 segmented, brown, hairy, tuff of scales are more on antenna, pilose; dark brown; pedicel 0.10 mm long, 0.09 mm broad, rounded, brown; flagellum 1.45 mm long, 15 segmented.

Flagellar Formula

3 L/W = 4.33, 9 L/W = 4.33, L 3/9 = 1, W 3/9 = 1, A = 2.66.

Thorax

1.40 mm long, 0.70 mm broad, brown, laterally compressed, dorsal brown stripes present; plurae uniformly not brown; scutum half moon shaped, brown smooth; scutellum (Figure 138) 0.40 mm long, 0.16 mm broad, triangular in lateral view, blackish brown; postnotum flat, not hairy; sternopleuron triangular, brown; metathoracic spiracle, rounded, yellow; lower mesepimeron bristles absent.

Forewing (Figure 139)

2.90 mm long, 0.60 mm broad, unspotted, transperant only costa is black in colour, scales 0.03 mm long, veins

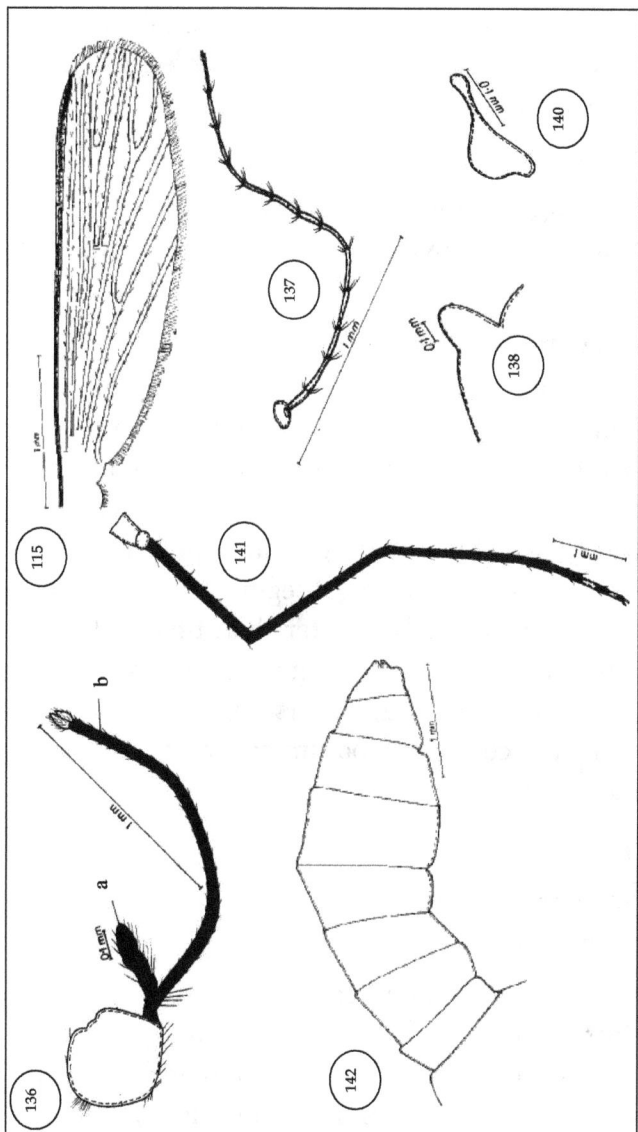

Plate 22–*Culex* (Culex) *mirjensis* sp. nov.

Figure 136: Head-Palpi (a) and Proboscis (b); **Figure 137:** Antenna; **Figure 138:** Scutellum; **Figure 139:** Forewing; **Figure 140:** Halter; **Figure 141:** Hind leg; **Figure 142:** Abdomen

sparsely scally; scales on 6th vein scanty, costa straight, dark blackish brown; subcosta straight, 2.75 mm long, reaching the costa; radius straight slightly curved, simple, without cross veins; cubitus bifurcated; anal vein curved, extend to wing margin, anal vein short.

Halter (Figure 140)

0.20 mm long, 0.15 mm broad, without scale, yellowish, triangular in lateral view, brown, expanded at tip; stalk faint yellow.

Hind Leg (Figure 141)

6.60 mm long, legs are not dark brown but yellowish brown, elongated longer than body; coxa 0.15 mm long, trochanter 0.10 mm long, yellowish, rounded; femur 1.60 mm long, cylindrical, yellowish, scally; tibia 2.10 mm long, slender, yellow, with two spurs, spurs are equal; tarsus 2.65 mm long, five segmented, 1st tarsal segment, 1.25 mm long, 1st tarsal segment distinctly shorter than tibia, 2nd tarsal segment 0.90 mm long, 3rd tarsal segment 0.25mm long, 4th tarsal segment 0.15mm long, 5th tarsal segment 0.10 mm long, claw simple curve; empodium and pulvillus, small; legs not dark brown.

Other Legs

Special marks : similar.

Abdomen (Figure 142)

1.90 mm long, 0.70 mm broad, dorsum of abdomen without pale band, tergite blackish brown, sternum yellowish brown; post genital plate 0.15 mm long, 0.09 mm broad, brown, hairy, anal cerci 0.09 mm long, 0.05 mm broad, hairy, brown, integument of plurae with dark spots.

Colour

Black	: Eyes.
Brown	: Vertex, clypeus, proboscis, antenna, thorax, legs.
Dark brown	: Head, abdomen.
Yellow	: Mandibles, maxillae, wing.
Yellowish brown	: Halter.
Male	: 2.80 mm long, smaller than female; antenna plumose, brushy, phytophagous.
Host	: Cattle.
Host Plant	: Unknown.
Holotype	: Female, India, Maharashtra, Miraj coll. Jagtap, M. B., 11-VI-2006; head, antenna, hind leg abdomen mounted on slide, labeled as above.
Paratype	: 37♂, 83 ♀, sex ratio (M:F) 1:1.50 coll. Jagtap, M. B., January 2006 to Oct. 2009.
Etymology	: The species name *Culex mirjensis* sp. nov. refers to the collection site of mosquito *i.e.* Miraj city, Maharashtra, India.

Distributional Record

2♂, 6♀, Tasgaon, 9-IV-2006; Vita, 10-VII-2006; 2♂, 9♀, Kagal, 12-VIII-2006; 3♂, 10♀, Jaysingpur, 12-XI-2006; 4♂, 9♀, 1♂, 2♀, Koregaon, 3-V-2007; 11♂, 21♀, Kolhapur, 8-VIII-2007; 9♂, 15♀, Miraj, 21-IX-2007; 5♂, 11♀, Kagal, 12-VII-2008.

Remark

According to the key of Barraud (1934) this species runs close to *Culex* (Barraudius) *modestus* Edward 1921 (Barraud, 1934; Sathe and Tingare, 2010) by having following characters.

1. Brown head and narrow scales are present.
2. 1st tarsal segment shorter than tibia.

However, it differs from above species by having following characters:

1. Legs are not dark brown but yellowish brown.
2. Wings are transparent only costa is black in colour.
3. Palpi not 1/6th of proboscis.
4. Hind femur shorter than tibia.
5. Proboscis without apical dark band.
6. Scutellum shape (triangular) and size (0.40 mm long and 0.16 mm broad).
7. Flagellar formula:

   3 L/W = 4.33, 9 L/W = 4.33, L 3/9 = 1, W 3/9 = 1, A = 2.66.

Culex **(Barraudius)** *satarensis* **sp. nov.**

Female (Figure 154)

4.95 mm long, 0.70 mm broad, small sized, dark brown, without ornamentation; head 0.50 mm long, 0.60 mm broad, blackish brown; antenna 2.10 mm long, brown; thorax 1.40 mm long, 0.70 mm broad, brown; wing 3 mm long, 1.70 mm broad, yellow; hind leg 7.40 mm long, brown; abdomen 3.10 mm long, 0.70 mm broad, blackish brown.

Head (Figure 143)

0.50 mm long, 0.60 mm broad, triangular, blackish brown, dorsal surface of head with narrow scales; compound eyes black, round; ocular space 0.15 mm, interocular distance 0.10 mm; vertex flat, brown, clypeus triangular, brown; proboscis 2.10 mm long, no pale ring on middle of proboscis, cylindrical, curved, brown, scally, labium 1.80 mm long, slender, curved, scally; labellum 0.30 mm long, densely scally, pointed; palpi 0.20 mm long and upwards, very shorter than proboscis, four segmented, hairy, brown, shorter than proboscis; mandible maxillae slender, longer, straight, yellow.

Antenna (Figure 144)

2.10 mm long, 15 segmented, dark brown, hairy, pilose; dark brown; pedicel 0.20 mm long, 0.08 mm broad, rounded, brown; flagellum 1.90 mm long, dark brown, 15 segmented, antenna without tuft of scale.

Flagellar Formula

3 L/W = 5.33, 9 L/W = 4.33, L 3/9 = 1.23, W 3/9 = 1, A = 2.97.

Thorax

1.40 mm long, 0.70 mm broad, brown, laterally compressed, dorsal brown stripes present; plurae uniformly not brown; scutum half moon shaped, brown smooth; scutellum (Figure 145) 0.36 mm long, 0.23 mm broad, yellowish scales on scutellum, triangular in lateral view, blackish brown; 3 black stripes on thorax; postnotum flat, not hairy; sternopleuron triangular, brown; metathoracic spiracle, rounded, yellow; lower mesepimeron bristles absent.

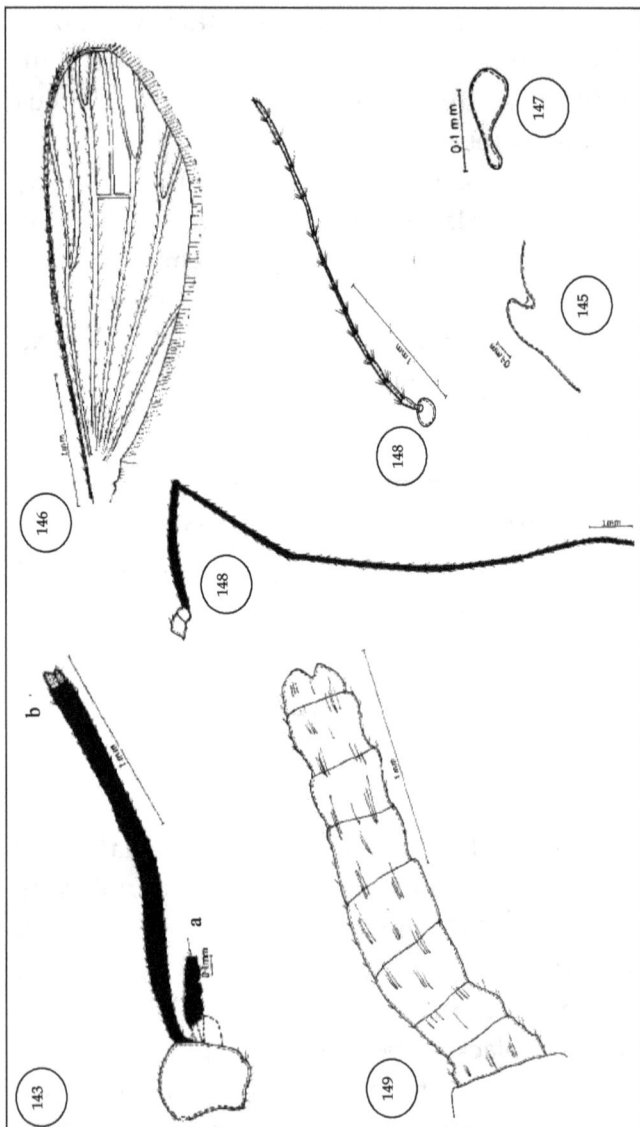

Plate 23–*Culex* (Culex) *satarensis* sp. nov.

Figure 143: Head-Palpi (a) and Proboscis (b); **Figure 144**: Antenna; **Figure 145**: Scutellum; **Figure 146**: Forewing; **Figure 147**: Halter; **Figure 148**: Hind leg; **Figure 149**: Abdomen

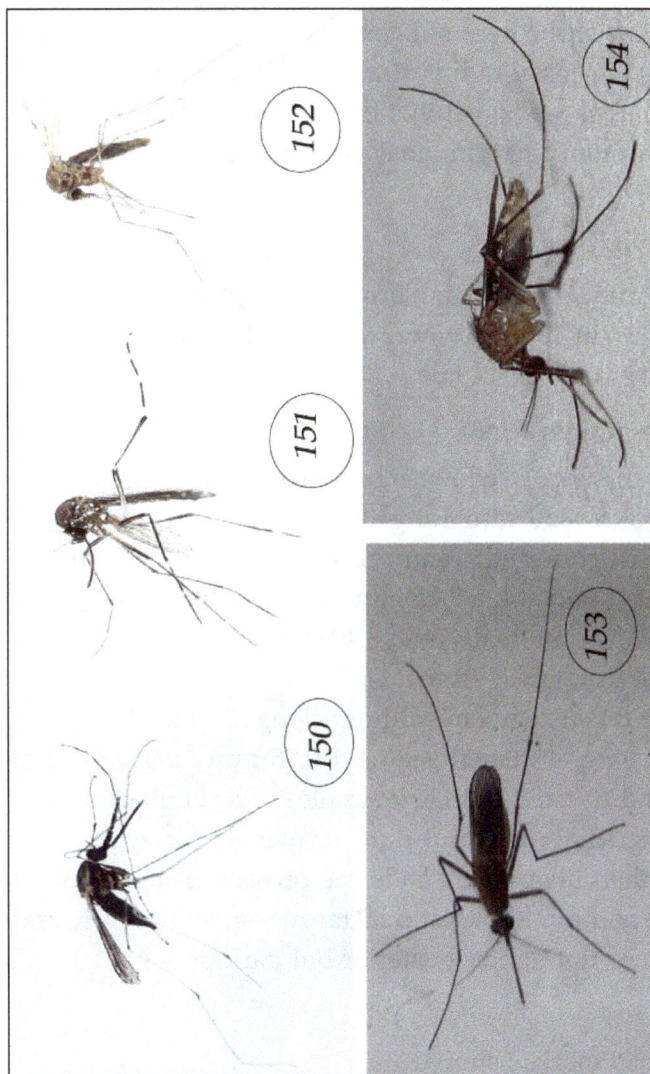

Plate 24—Figure 150: *Culex* (Culex) *quinquefasciatus* sp. nov.; **Figure 151:** *Culex* (Culex) *malhari* sp. nov.; **Figure 152:** *Culex* (Culex) *malkapuri* sp. nov.; **Figure 153:** *Culex* (Barraudius) *mirjensis* sp. nov.; **Figure 154:** *Culex* (Barraudius) *satarensis* sp. nov.

Forewing (Figure 146)

3 mm long, 1.70 mm broad, unspotted, transperant, scales 0.03 mm long, veins sparsely scally; costa straight, dark blackish brown; subcosta straight, 2.60 mm long, reaching the costa; radius straight slightly curved, simple, without cross veins; cubitus bifurcated; anal vein curved, extend to wing margin, anal vein short, scanty scales on vein 6.

Halter (Figure 147)

0.25 mm long, 0.15 mm broad, without scale, yellowish, triangular in lateral view, dumb-bell shaped, brown, expanded at tip; stalk faint yellow.

Hind Leg (Figure 148)

7.40 mm long, yellowish brown, elongated longer than body; coxa 0.20 mm long; trochanter 0.10 mm long, yellowish, rounded; femur 1.55 mm long, cylindrical, yellowish, scally; tibia 2.15 mm long, slender, yellow, with two spurs, spurs equal; tarsus 3.40 mm long, five segmented, Pale bands on tarsal segments; 1st tarsal segment, 1.60 mm long, 2nd tarsal segment 1.10 mm long, 3rd tarsal segment 0.35mm long, 4th tarsal segment 0.25 mm long, 5th tarsal segment 0.10 mm long, claw simple curve; empodium and pulvillus, small; legs not dark brown, 1st tarsal segment shorter than tibia. Pale bands are present on the joints of hind leg and foreleg, 4th and 5th tarsal segments of fore and hind femur not lighter posterior but dark brown.

Other Legs

Special marks : similar.

Abdomen (Figure 149)

3.10 mm long, 0.70 mm broad, dorsum of abdomen with pale bands, no continuous pale boarder on either side, tergite blackish brown, sternum yellowish brown; post genital plate 0.15 mm long, 0.08 mm broad, brown, hairy, anal cerci 0.08 mm long, 0.06 mm broad, hairy, brown, integument of plurae with dark spots.

Colour

Black	: Eyes.
Brown	: Vertex, clypeus, proboscis, antenna, thorax, legs.
Dark brown	: Head, abdomen.
Yellow	: Mandibles, maxillae, wing.
Yellowish brown	: Halter.
Male	: 3.80 mm long, smaller than female; antenna plumose, brushy, phytophagous.
Host	: Cattle.
Host Plant	: Unknown.
Holotype	: Female, India, Maharashtra, Satara coll. Jagtap, M. B., 27-VI-2009; head, antenna, hind leg abdomen mounted and pinnedon slide, labelled as above.
Paratype	: 63♂, 162 ♀, sex ratio (M:F) 1:2.57 coll. Jagtap, M. B., Jun 2006 to Nov. 2009.
Etymology	: The species name *Culex* (Barraudius) *satarensis* sp. nov. refers to the collection site of mosquito *i.e.* Satara city, district Satara, Maharashtra, India.

Distributional Record

8♂, 18♀, Miraj, 11-VI-2006; 5♂, 13♀, Jaysingpur, 12-XI-2006; 4♂, 11 ♀, Miraj, 14-III-2007; 5 ♂, 12♀, Pune, 25-VII-2007; 5♂, 17♀, Kolhapur, 8-VIII-2007; 9♂, 21♀, Wai, 9-II-2008; 2♂, 9 ♀, Jaysingpur, 21-IX-2008; 3♂, 9 ♀, Satara, 13-XII-2008, 5 ♂, 13♀, Satara, 27-VI-2009; 7♂, 18♀, Junner, 22-VIII-2009; 7♂, 14 ♀, Baramati, 28-XI-2009.

Remark

According to the key of Barraud (1934) this species runs close to *Culex* (Barraudius) *modestus* Ficalbi (Barraud, 1934; Sathe and Tingare, 2010) by having following characters.

1. Head with brown narrow scales.
2. Colour of palpi, proboscis, legs and antenna dark brown.

However, it differs from above species by having following characters:

1. Wings are not dark only costa is black in colour.
2. Pale bands are present on the joints of hind leg and foreleg, 4th and 5th tarsal segment.
3. Scutellar scales yellowish.
4. 3 Black stripes on thorax.
5. Fore and hind femur not lighter posterior but dark brown.
6. Abdomen without continues pale boarder on either side
7. Antenna without tuft of scales.
8. Scutellum shape (triangular) and size (0.36 mm long and 0.23 mm broad).

9. Flagellar formula:

3 L/W = 5.33, 9 L/W = 4.33, L 3/9 = 1.23, W 3/9 = 1, A = 2.97.

Anopheles (Cellia) *subpictus*

Female

4.16 mm long, 0.95 mm broad; head 0.76 mm long, 0.65 mm broad, blackish brown; antenna 1.85 mm long, brownish; thorax 1.65 mm long, 0.65 broad, blackish brown; forewing 3.45 mm long, 0.85 mm broad, marked spotted wings, yellowish; hind leg 6.80 mm long, yellowish; abdomen 1.75 mm long, 0.95 mm broad, yellowish brown.

Head (Figure 155)

0.76 mm long, 0.65 mm broad, blackish brown, globular, with narrow scales; compound eyes black, rounded, ocular space 0.30 mm, interocular distance 0.22 mm long; vertex smooth, dark brown; proboscis 2.25 mm long, blackish, scally, cylindrical; labium 0.20 brown, yellowish, scally; labellum 2.05 mm long, brownish, scally; palpi 2.00 mm long, 5 segmented, slender, as long as proboscis, densely scally, palpi completely dark but small bands at joints.

Antenna (Figure 156)

1.85 mm long, 15 segmented, hairy, brownish, pilose; scape 0.06 mm long, 0.10 mm broad, brownish; pedicel 0.30 mm long, 0.15 mm broad, yellowish brown; flagellum 1.40 mm long, 13 segmented and 3 pale bands. Apical pale band nearly equal to the preapical dark band.

Flagellar Formula

1 L/W = 3.8, 13 L/W = 2.8, L1/13 = 1, W1/13 = 0.8, A = 2.1.

Thorax

1.65 mm long, 0.65 mm broad, blackish brown, undifferentiated, laterally compressed, lyre shaped, scutum black, without scales, white spots on thorax; scutellum 0.6 mm long, 0.40 mm broad; rounded, globular and opaque, brownish; sternopleuron and mesepimeron triangular, brownish.

Forewing (Figure 157)

3.45 mm long, 0.85 mm broad, wing with pale marking wide dark pale band, golden, scales present on veins; subcosta straight, 2.55 mm long, reaching costa, Inner costa interrupted; media straight, 2.60 mm long; radius straight, slightly curved at apex; cubitus bifurcated; Dark pale band on forewing is 0.9 mm long and vein no. 3 is dark. Fringe spot on all the veins.

Halter (Figure 158)

0.42 mm long, 0.09 mm broad, brownish tubular shaped, without scales or hairs, expanded at tip.

Hind Leg (Figure 159)

6.80 mm long, yellowish, longer than body, coxa 0.30 mm long 0.15 mm broad, yellowish, trochanter 0.15 mm long, 0.13 mm broad, rounded, yellowish; femur 1.50 mm long, cylindrical, yellowish, covered with scales; femora not speckled, mid femur with large pale spot; tibia 1.80 mm long, yellowish, not speckled, hind femur with not distinct knee spot, tibia bristles present in between joints; tarsus 3.05 mm long, yellowish, scally, five segmented, 1st tarsal segment 1.05 mm long, 2nd tarsal segment 0.80 mm long, 3rd tarsal segment 0.65 mm long, 4th tarsal segment 0.30 mm long, 5th tarsal segment 0.15 mm long, pretarsus longer than other. Fore leg tarsomeres with broad bands.

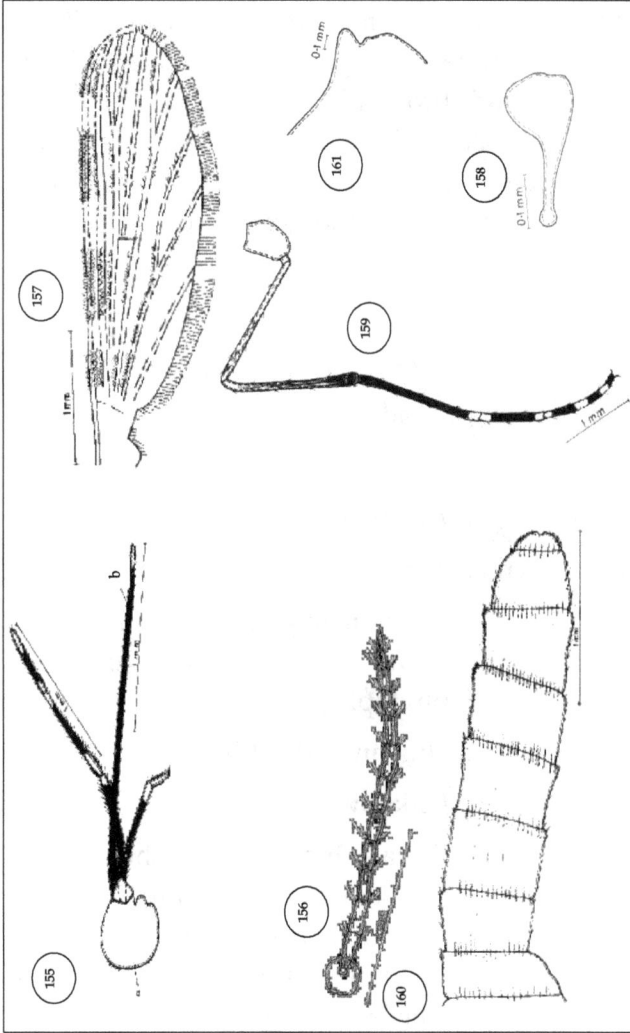

Plate 25—*Anopheles* (Cellia) *subpictus*

Figure 155: Head-Palpi (1a) and Proboscis (1b); **Figure 156**: Antenna; **Figure 157**: Forewing; **Figure 158**: Halter; **Figure 159**: Hind leg; **Figure 160**: Abdomen; **Figure 161**: Scutellum

Other Legs

Special marks: similar.

Abdomen (Figure 160)

1.75 mm long, 0.95 mm broad, reddish brown, without banded, dorsal plate reddish; post genital plate 0.85 mm long, 0.08 mm broad, brownish, hairy, longer than anal cerci, post genital plate and anal cerci in right angle; anal cerci 0.05 mm long, 0.03 mm broad, densely hairy, brown, Last 4 segments very narrow and golden colour hairs are present.

Colour

Black	: Eyes.
Blackish brown	: Head.
Yellowish	: Proboscis, labium, wing, thorax.
Brownish	: Antenna, halter, legs.
Reddish brown	: Abdomen.
Male	: 3.15 mm long, slender, smaller than female; antenna plumose, phytophagus.
Host	: Human and cattle.
Host plant	: Unknown.
Holotype	: Female, India, Maharashtra, Satara coll. Jagtap, M.B. 27-VII-2009 head, antenna, hind leg, wing and abdomen mounted and pinned, labelled as above.
Paratype	: 76 ♂, 195 ♀, sex ratio (M:F) 1:2.56 coll. Jagtap, M.B. Mar. 2006 to Dec. 2009.

Distributional Record

3 ♂ 7 ♀ Saswad, 12-III-2006; 5♂ 11 ♀ Vita, 10-VI-2006; 8♂, 18♀, Miraj, 11-VI-2006; 5♂, 13♀, Jaysingpur, 12-XI-2006; 4♂, 11 ♀, Miraj, 14-III-2007; 5 ♂, 12♀, Pune, 25-VII-2007; 5♂, 17♀, Kolhapur, 8-VIII-2007; 9♂, 21♀, Wai, 9-II-2008; 2 ♂, 7 ♀ Medha 20-VII-2008; 2♂, 9 ♀, Jaysingpur, 21-IX-2008; 3♂, 9 ♀, Satara, 13-XII-2008, 5 ♂, 13♀, Satara, 27-VI-2009; 6 ♂ 15 ♀ Ajara, 25-VII-2009; 7♂, 18♀, Junner, 22-VIII-2009; 7♂, 14 ♀, Baramati, 28-XI-2009.

Remark

According to the key of Rao (1984) this is *Anopheles* (Anopheles) *subpictus* Grassi 1899 by following characters.

1. Apical pale band nearly equal to the preapical dark band
2. Fore leg tarsomeres with broad bands.
3. Fringe spot on all the veins.

However, some following additional characters have been found:

1. Dark pale band on forewing is 0.9 mm long and vein no. 3 is dark.
2. Scutellum globular and opaque 0.6mm long and 0.40 mm broad.
3. Flagellar formula:
   1 L/W = 3.8, 13 L/W = 2.8, L1/13 = 1, W1/13 = 0.8, A = 2.1.

Anopheles (Cellia) *fluviatilis*

Female

4.10 mm long, 0.90 mm broad; head 0.70 mm long, 0.60 mm broad, blackish brown; antenna 1.80 mm long, brownish; thorax 1.70 mm long, 0.65 broad, blackish brown; forewing 3.30 mm long, 0.85 mm broad, marked spotted wings, yellowish; hind leg 6.80 mm long, yellowish; abdomen 1.75 mm long, 0.90 mm broad, yellowish brown.

Head (Figure 162)

0.70 mm long, 0.60 mm broad, blackish brown, globular, with narrow scales; compound eyes black, rounded, ocular space 0.33 mm, interocular distance 0.21mm long; vertex smooth, dark brown; proboscis 2.35 mm long, blackish, scally, cylindrical; labium 0.25 brown, yellowish, scally; labellum 2.10 mm long, brownish, scally; palpi 2.00 mm long, 5 segmented, slender, as long as proboscis, densely scally, palpi completely dark but small bands at joints.

Antenna (Figure 163)

1.80 mm long, 15 segmented, hairy, brownish, pilose; scape 0.06 mm long, 0.10 mm broad, brownish; pedicel 0.30 mm long, 0.15 mm broad, yellowish brown; flagellum 1.35 mm long, 13 segmented and 3 pale bands. Apical pale band nearly equal to the preapical dark band.

Flagellar Formula

1 L/W = 3.6, 13 L/W = 2.5, L1/13 = 1, W1/13 = 0.7, A = 1.95.

Thorax

1.70 mm long, 0.65 mm broad, blackish brown, undifferentiated, laterally compressed, lyre shaped, scutum black, without scales, white spots on thorax; scutellum 0.5 mm long, 0.43 mm broad; rounded, globular and opaque, brownish; sternopleuron and mesepimeron triangular, brownish.

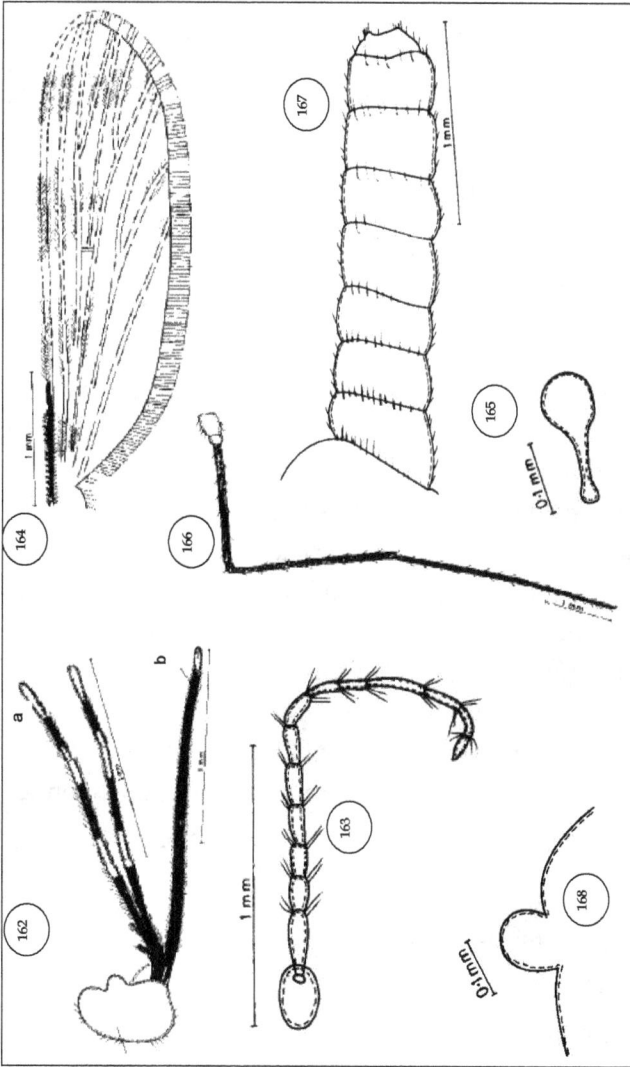

Plate 26—*Anopheles* (Cellia) *fluviatilis*

Figure 162: Head-Palpi (1a) and Proboscis (1b); **Figure 163**: Antenna; **Figure 164**: Forewing; **Figure 165**: Halter; **Figure 166**: Hind leg; **Figure 167**: Abdomen; **Figure 168**: Scutellum

Forewing (Figure 164)

3.30 mm long, 0.85 mm broad, wing with pale marking wide dark pale band, golden, scales present on veins; subcosta straight, 2.50 mm long, reaching costa, Inner costa interrupted; inner costa mainly dark media straight, 2.65 mm long; radius straight, slightly curved at apex; cubitus bifurcated; Vein 3 mainly pale.

Halter (Figure 165)

0.43 mm long, 0.08 mm broad, brownish tubular shaped, without scales or hairs, expanded at tip.

Hind Leg (Figure 166)

6.80 mm long, yellowish, longer than body, coxa 0.25 mm long 0.10 mm broad, yellowish, trochanter 0.10 mm long, 0.13 mm broad, rounded, yellowish; femur 1.50 mm long, cylindrical, yellowish, covered with scales; femora not speckled, mid femur with large pale spot; tibia 1.80 mm long, yellowish, not speckled, hind femur with not distinct knee spot, tibia bristles present in between joints; tarsus 3.10 mm long, yellowish, scally, five segmented, 1st tarsal segment 1.10 mm long, 2nd tarsal segment 0.85 mm long, 3rd tarsal segment 0.65 mm long, 4th tarsal segment 0.35 mm long, 5th tarsal segment 0.15 mm long, pretarsus longer than other. Tarsomeres without bands.

Other Legs

Special marks: similar.

Abdomen (Figure 167)

1.75 mm long, 0.90 mm broad, raddish brown, without banded, dorsal plate reddish; post genital plate 0.80 mm long, 0.07 mm broad, brownish, hairy, longer than anal cerci, post genital plate and anal cerci in right angle; anal

cerci 0.05 mm long, 0.03 mm broad, densely hairy, brown,
Last 4 segments very narrow and golden colour hairs are
present.

Colour

Black	: Eyes.
Blackish brown	: Head.
Yellowish	: Proboscis, labium, wing, thorax.
Brownish	: Antenna, halter, legs.
Reddish brown	: Abdomen.
Male	: 3.10 mm long, slender, smaller than female; antenna plumose, phytophagus.
Host	: Human and cattle.
Host plant	: Unknown.
Holotype	: Female, India, Maharashtra, Pune coll. Jagtap, M.B. 25-VII-2007 head, antenna, hind leg, wing and abdomen mounted and pinned, labelled as above.
Paratype	: 26 ♂, 62 ♀, sex ratio (M:F) 1:2.38 coll. Jagtap, M.B. July 2007 to Dec. 2009.

Distributional Record

3 ♂, 10 ♀, Pune, 25-VII-2007; 1♂, 3 ♀, Kolhapur, 8-VIII-2007; 3♂, 5 ♀, Wai, 9-II-2008; 2 ♂, 6 ♀ Medha 20-VII-2008; 2♂, 9 ♀, Satara, 13-XII-2008, 6 ♂ 15 ♀ Ajara, 25-VII-2009; 5♂, 8♀, Junner, 22-VIII-2009; 4♂, 6 ♀, Baramati, 28-XI-2009.

Remark

According to the key of Rao (1984) this species is *Anopheles* (Anopheles) *fluviatilis* James 1902 by following characters.

1. Apical pale band nearly equal to the preapical dark band.
2. Tarsomeres without bands.
3. Vein 3 mainly pale.
4. Inner costa mainly dark.

However, some following additional characters have been found:

1. Scutellum globular and opaque 0.5 mm long, 0.43 mm broad.
2. Flagellar formula:
   1 L/W = 3.6, 13 L/W = 2.5, L1/13 = 1, W1/13 = 0.7, A = 1.95.

Culex (Culex) *vishnui*

Female

4.25 mm long, 0.75 mm broad, small sized, dark brown, without ornamentation; head 0.55 mm long, 0.60 mm broad, blackish brown; antenna 2.15 mm long, brown; thorax 1.45 mm long, 0.75 mm broad, brown; wing 3 mm long, 1.70 mm broad, yellow; hind leg 7.40 mm long, brown; abdomen 2.25 mm long, 0.70 mm broad, blackish brown.

Head (Figure 169)

0.55 mm long, 0.60 mm broad, triangular, blackish brown, dorsal surface of head with narrow scales and upright scales on vertex; compound eyes black, round;

ocular space 0.15 mm, interocular distance 0.10 mm; vertex flat, brown, clypeus triangular, brown; proboscis 2.30 mm long, pale band on middle of proboscis, cylindrical, curved, dark brown, scally, labium 1.80 mm long, slender, curved, scally; labellum 0.30 mm long, densely scally, pointed; palpi 0.20 mm long dark brown and upwards, very shorter than proboscis, four segmented, hairy, brown, shorter than proboscis; mandible maxillae slender, longer, straight, yellow.

Antenna (Figure 170)

2.15 mm long, 15 segmented, dark brown, hairy, pilose; dark brown; pedicel 0.20 mm long, 0.08 mm broad, rounded, brown; flagellum 1.95 mm long, dark brown, 15 segmented, antenna without tuft of scale.

Flagellar Formula

3 L/W = 5.40, 9 L/W = 4.20, L 3/9 = 1.13, W 3/9 = 1, A = 2.93.

Thorax

1.45 mm long, 0.75 mm broad, brown, laterally compressed, dorsal brown stripes present; plurae uniformly not brown, scutum half moon shaped, brown smooth; mesonotum clothed with golden brown narrow scales, scutellum 0.35 mm long, 0.22 mm broad, triangular in lateral view, blackish brown; 3 black stripes on thorax; postnotum flat, not hairy; sternopleuron triangular, brown; metathoracic spiracle, rounded, yellow; lower mesepimeron bristles absent.

Forewing (Figure 171)

3.1 mm long, 1.70 mm broad, unspotted, with dark scales, scales 0.03 mm long, veins sparsely scally; costa

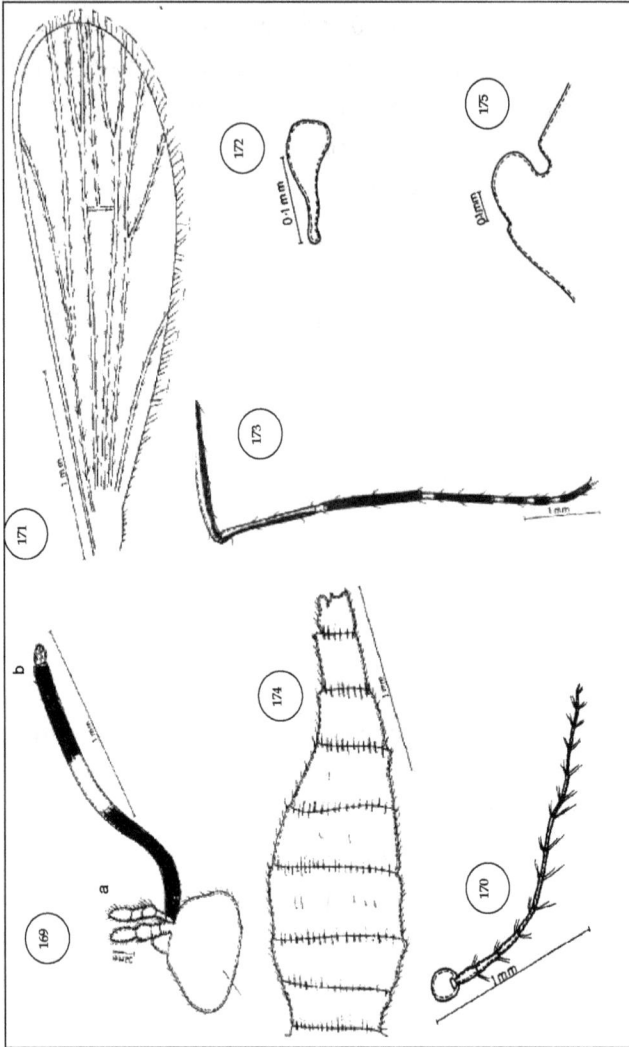

Plate 27–*Culex* (Culex) *vishnui*

Figure 169: Head-Palpi (1a) and Proboscis (1b); **Figure 170**: Antenna; **Figure 171**: Forewing; **Figure 172**: Halter; **Figure 173**: Hind leg; **Figure 174**: Abdomen; **Figure 175**: Scutellum

straight, dark blackish brown; subcosta straight, 2.60 mm long, reaching the costa; radius straight slightly curved, simple, without cross veins; cubitus bifurcated; anal vein curved, extend to wing margin, anal vein short, scanty scales on vein 6.

Halter (Figure 172)

0.22 mm long, 0.15 mm broad, without scale, yellowish, triangular in lateral view, dumb-bell shape, brown, expanded at tip; stalk faint yellow.

Hind Leg (Figure 173)

7.20 mm long, dark brown, elongated longer than body; coxa 0.15 mm long, trochanter 0.10 mm long, yellowish, rounded; femur 1.50 mm long and dark brown, cylindrical, yellowish, scally; tibia 2.10 mm long, slender, yellow, with two spurs, spurs equal; tarsus 3.35 mm long, five segmented, Pale bands on tarsal segments; 1st tarsal segment, 1.55 mm long, 2nd tarsal segment 1.10 mm long, 3rd tarsal segment 0.35mm long, 4th tarsal segment 0.25mm long, 5th tarsal segment 0.10 mm long, claw simple curve; empodium and pulvillus, small; legs not dark brown, 1st tarsal segment shorter than tibia.

Other Legs

Special marks : similar.

Abdomen (Figure 174)

2.25 mm long, 0.70 mm broad, dorsum of abdomen with pale bands, tergite dark brown with well marked basal bands, sternum yellowish brown; post genital plate 0.17 mm long, 0.08 mm broad, brown, hairy, anal cerci 0.08 mm long, 0.06 mm broad, hairy, brown, integument of plurae with dark spots.

Colour

Black	: Eyes
Brown	: Vertex, clypeus, antenna, thorax.
Dark brown	: Head, proboscis, palpi, legs. abdomen.
Yellow	: Mandibles, maxillae, wing.
Yellowish brown	: Halter.
Male	: 3.10 mm long, smaller than female; antenna plumose, brushy, phytophagous.
Host	: Cattle.
Host Plant	: Unknown.
Holotype	: Female, India, Maharashtra, Ajara coll. Jagtap, M. B., 25-VIII -2009; head, antenna, hind leg abdomen mounted and pinned on slide, labelled as above.
Paratype	: 13 ♂, 29 ♀, sex ratio (M:F) 1:2.30 coll. Jagtap, M. B., Aug 2007 to July 2009.

Distributional Record

1♂, 3 ♀, Kolhapur, 8-VIII-2007; 2 ♂, 4 ♀, Jaysingpur, 21-IX-2008; 3♂, 5 ♀, Junner, 22-VIII-2009; 1♂, 3 ♀, Baramati, 28-XI-2009. 2 ♂, 5 ♀ Medha 20-VII-2008; 4 ♂ 9 ♀ Ajara, 25-VII-2009.

Remark

According to the key of Barraud (1934) this species is *Culex* (Culex) *vishnui* Theobald 1901 by having following characters.

1. Pale band on middle of proboscis.
2. Narrow and upright pale brown scales on vertex.
3. Palpi dark brown.
4. Tarsi with distinct pale ring.
5. Tergite dark brown with well marked basal bands.
6. Scutellum shape (triangular) and size (0.35 mm long and 0.22 mm broad).
7. Flagellar formula:

   3 L/W = 5.40, 9 L/W = 4.20, L 3/9 = 1.13, W 3/9 = 1, A = 2.93.

Chapter 5
Seasonal Abundance and Distribution of Mosquitoes

Introduction

Entomological information needs to be upgraded to monitor the impact of rapidly changing ecological conditions such as deforestation, population movement and developmental activities on mosquito distribution and vector bionomics (Rahman *et al.,* 1977). Seasonal abundance of mosquitoes may vary spatially. Sampling of mosquito population is an important task, which estimate the number of species presents in a target area. Patterns of seasonal abundance of certain mosquito species are correlated to proliferation of its breeding habitats during rainy season and its scarcity during dry season (White, 1974). Various populations have specific characteristic features which facilitate the formation of epidemiological characteristics of vector borne diseases (Kondrashin and Kalra, 1987).

Rapidly changing environment brings about frequent changes in vector behavior, which affects the vector bionomics (Prakash *et al.*, 1998).

Global warming is reshaping the ecology of many medically important arthropod vectors. Warmer temperatures have been shown to directly increase mosquito biting and pathogen transmission.

Review of Literature

Review of literature indicates that several workers (Senior White, 1937; Senior White *et al.*, 1943; Foot and Cook, 1959; Nagpal *et al.*, 1983; Das *et al.*, 1984; Rao, 1984; Nagpal and Sharma, 1987; Nagpal and Sharma, 1995; Reuben *et al.*, 1992; Rajavel *et al.*, 2000; Sathe and Girhe, 2001a, 2001b, 2001c; Murty *et al.*, 2002; Kanojia *et al.*, 2003; Sharma *et al.*, 2005; Joshi *et al.*, 2005; Tilak *et al.*, 2006; Pemola and Jauhari, 2006; Malarial Research Centre, 2006; Baruah *et al.*, 2007; Jagtap and Sathe, 2008a; Jagtap and Sathe, 2008b; Jagtap and Sathe, 2008c; Sathe and Jagtap, 2009; Jagtap and Sathe, 2009; Sathe and Tingare, 2010; Sathe and Jagtap, 2010 etc.) attempted abundance of mosquitoes from India. The present work is precise attempt on the seasonal abundance of mosquitoes and will add great relevance in solving cases of mosquito borne diseases in the region.

Materials and Methods

The survey of mosquitoes was made from Western Maharashtra (districts Pune, Satara, Kolhapur and Sangli) (Figure 2) from 2005 to 2010. A large number of specimen were collected by visiting various places of Western Maharashtra namely, districts Pune (Baramati, Pune, Bhor, Saswad, Haveli, Junner), Satara (Medha, Wai,

Mahabaleshwar, Satara, Patan, Mhaswad, Koregaon), Kolhapur (Ajra, Malkapur, Kagal, Kolhapur and Jaysingpur) and Sangli (Miraj, Vita, Tasgaon, Shirala and Jath) at 15 days interval.

The mosquito surveillance was carried out indoor as well as outdoor. Mosquito surveillance started in early in the morning from 6.15 am or in evening after 6.30 pm. Mosquitoes were collected by suction tube were transported in to test tubes for further identification. Larvae were collected with the help of ladle and dropper by one-man one-hour search. Larvae and pupae of mosquitoes from natural habitats from selected spots have also been collected and reared in the laboratory for their adult formation. The specimen collected during study period were identified by consulting Christopher (1933), Barraud (1934), Horsfall (1955), Rao (1984), Nagpal and Sharma (1994), Sathe and Girhe (2002) and Sathe and Tingare (2010). Distribution records of the specimens have been made by visiting, collecting and identifying the species from study spots of Western Maharashtra.

Restuls

Results are recorded in Tables 1 to 3.

The observations on seasonal abundance of mosquito species belonging to genera *Anopheles, Culex* and *Aedes* indicates that out of 31 species 10 species were rare and 21 species were common in the Western Maharashtra.

Among the total 3362 mosquito species collected the highest contribution was from *Anopheles* (45.33 per cent) followed by *Culex* (35.01 per cent), *Armiger* (10.71 per cent), and *Aedes* (8.95 per cent). The remaining 26 species were

Table 1: Check List of Mosquitoes from Southern Maharashtra

Sl.No.	Mosquito Species	Abundance	Citations
FAMILY - CULICIDAE SUB FAMILY - ANOPHELINAE GENUS - ANOPHELES			
1.	*Anopheles culicifacies* Giles	Common	1901b. *Ent. Mon. Mag.* 37 : 196-198.
2.	*Anopheles stephensi* Liston	Common	1901. *Indian Med. Gat.* 36 : 361-366.
3.	*Anopheles annularis* Vander Wulp.	Common	1884. *Notes Leyden Mus.* 6 : 248-256.
4.	*Anopheles subpictus* Grassi	Common	1899. *Indian Entomologist* 34 : 192-197.
5.	*Anopheles turkhudi* Liston	Rare	1901. *Indian Med. Gat.* 36 : 441-443.
6.	*Anopheles compestris* Reid.	Common	1962. *Notes Leyden Mus.* 6 : 248-256.
7.	*Anopheles culiciformis* Cogill	Common	1903. *J. Bombay nat. Hist. Soc.* 15 : 327-336, 1 pl.
8.	*Anopheles jeyporeiensis* James,	Rare	1902. *Sci. Mem. Med. Sanit. Dept. India (N.S.)* No.2, 106 pp.
9.	*Anopheles karwari* James	Rare	1902. *Sci. Mem. Med. Sanit. Dept. India (N.S.)* No.2, 106 pp.
10.	*Anopheles maculatus* Theobald,	Rare	1901.*A mon. of Culicidae or Mosq,* 1:171-174.
11.	*Anopheles vagus* Doenitz.	Rare	*1902. Zeit. Fur Hyg. Und Infek., 41: 15-88.*
12.	*Anopheles mahabaleshwari* sp. nov.	Common	
13.	*Anopheles waii* sp. nov.	Common	
14.	*Anopheles karveeri* sp. nov.	Common	

Contd...

Table 1–*Contd...*

Sl.No.	Mosquito Species	Abundance	Citations
15.	*Anopheles krishnai* sp. nov.	Common	
16.	*Anopheles kolhapuri* sp. nov.	Common	
SUB FAMILY - CULICINAE GENUS - CULEX			
17.	*Culex epidesmus* Theobald	Rare	1910a. *Royal Society* 12 pp. *British Museum (Nat. His.)*.
18.	*Culex tritaeniorhynchus* Giles	Common	1901a. *J. Bombay Soc.* 13 : 592-610, pls. A and B.
19.	*Culex vishnui* Theobald	Common	1910. *Rec. Indian Mus.* 4 : 1-33, 3 pls.
20.	*Culex quinquefasciatus* Say	Common	1823. *J. Acad. Nat. Sci. Philad* 3 : 9-54.
21.	*Culex fuscocephala* Theobald	Rare	1907. M. C. iv. P. 420.
22.	*Culex malhari* sp. nov.	Common	
23.	*Culex malkapuri* sp. nov.	Common	
24.	*Culex satarensis* sp.nov	Common	
25.	*Culex mirjensis* sp. nov	Common	
SUB FAMILY - CULICINAE GENUS – ARMIGERS			
26.	*Armiger (Armiger) subalbatus* Coquillett.	Common	1898. *Rec. Indian Mus.* 4 : 1-33, 3 pls.

Contd...

Table 1–*Contd*...

Sl.No.	Mosquito Species	Abundance	Citations
SUB FAMILY - CULICINAE GENUS - AEDES			
27.	*Aedes aegypti* Linnaeus	Common	1762. Zweyter Theil, ent. Besc. varschiedener wichtiger Naturalien pp. 267-606.
28.	*Aedes albopictus* Skuse	Common	1894. *Indian Mus. Notes.* 3, No. 5, p. 20.
29.	*Aedes vittatus* Bigot	Rare	1861. *Ann. Soc. ent. Fr.* (4) 1 : 227-229.
30.	*Aedes* (Mucidus) *sathei* sp. nov.	Rare	
31.	*Aedes* (*Finalaya*) *rajashri*. sp. nov.	Rare	

Table 2: Seasonal Abundance of Mosquitoes from Selected Spots

Sl.No.	Species	Monsoon (Jun -Sept)		Post Monsoon (Oct -Jan)		Pre-monsoon (Feb -May)		Total	
		Total Mosquito Catch	Density (DP MH)	Total Mosquito Catch	Density (DP MH)	Total Mosquito Catch	Density (DP MH)	Total Mosquito Catch	Density (DP MH)
1.	Anopheles (Anopheles) *culiciformis* Cogill, 1903.	9	3.6	5	2	3	1.2	17	6.8
2.	Anopheles (Anopheles) *kolhapuri* sp. nov.	17	6.8	7	2.8	9	3.6	33	13.2
3.	Anopheles (Anopheles) *compestris* Reid, 1962.	26	10.4	20	8	1	0.4	47	18.8
4.	Anopheles (Cellia) *annularis* Vander Wulp, 1884.	11	4.4	27	10.8	5	2	43	17.2
5.	Anopheles (Cellia) *turkhudi* Liston, 1901	9	3.6	3	1.2	0	0	12	4.8
6.	Anopheles (Cellia) *stephensii* Liston, 1901.	9	3.6	16	6.4	29	11.6	54	21.6
7.	Anopheles (Cellia) *culicifacies* Giles, 1901.	242	96.8	76	30.4	193	77.2	511	204.4
8.	Anopheles (Cellia) *jeyporeiensis* James, 1902.	15	6	1	0.4	0	0	16	6.4
9.	Anopheles (Cellia) *karwari* (James), 1902.	11	4.4	6	2.4	1	0.4	18	7.2
10.	Anopheles (Cellia) *maculatus* Theobald, 1901.	2	0.8	4	1.6	3	1.2	9	3.6
11.	Anopheles (Cellia) *vagus* Doenitz, 1902.	2	0.8	1	0.4	1	0.4	4	1.6
12.	Anopheles (Cellia) *subpictus* Grassi, 1899.	92	36.8	102	40.8	39	15.6	233	93.2

Contd...

Table 2–*Contd...*

Sl.No.	Species	Monsoon (Jun-Sept)		Post Monsoon (Oct-Jan)		Pre-monsoon (Feb-May)		Total	
		Mosquito Catch	Total Density (DP MH)	Mosquito Catch	Total Density (DP MH)	Mosquito Catch	Total Density (DP MH)	Mosquito Catch	Total Density (DP MH)
13.	*Anopheles* (*Cellia*) *mahabaleshwari* sp. nov.	19	7.6	27	10.8	20	8	66	26.4
14.	*Anopheles* (*Cellia*) *waii* sp. nov.	21	8.4	22	8.8	14	5.6	57	22.8
15.	*Anopheles* (*Cellia*) *karveeri* sp. nov.	35	14	7	2.8	12	4.8	54	21.6
16.	*Anopheles* (*Cellia*) *krishnai* sp. nov.	203	81.2	97	38.8	50	20	350	140
	Total	**723**	**289.2**	**421**	**168.4**	**380**	**152**	**1524**	**609.6**
17.	*Aedes* (*Stegomyia*) *vittatus* (Bigot), 1861.	11	4.4	3	1.2	2	0.8	16	6.4
18.	*Aedes* (*Stegomyia*) *aegypti* (Linnaeus), 1762	89	35.6	31	12.4	23	9.2	128	51.2
19.	*Aedes* (*Stegomyia*) *albopictus* (Skuse), 1894	77	30.8	25	10	34	13.6	136	54.4
20.	*Aedes* (*Mucidus*) *sathei* sp. nov.	0	0	3	1.2	1	0.4	4	1.6
21.	*Aedes* (*Finalaya*) *rajashri* sp. nov.	10	4	7	2.8	0	0	17	6.8
	Total	**187**	**74.8**	**69**	**27.6**	**60**	**24**	**301**	**120.4**

Contd...

Table 2—Contd...

Sl.No.	Species	Monsoon (Jun–Sept)		Post Monsoon (Oct–Jan)		Pre-monsoon (Feb–May)		Total	
		Total Mosquito Catch	Density (DP MH)	Total Mosquito Catch	Density (DP MH)	Total Mosquito Catch	Density (DP MH)	Total Mosquito Catch	Density (DP MH)
22.	Culex (Culex) quinquieasciatus Say, 1823.	251	100.4	95	38	158	63.2	504	201.6
23.	Culex (Culex) epidesmus (Theobald), 1910.	22	8.8	9	3.6	7	2.8	38	15.2
24.	Culex (Culex) vishnui Theobald, 1901.	6	2.4	26	10.4	2	0.8	34	13.6
25.	Culex (Culex) fuscocephala Theobald, 1907.	8	3.2	4	1.6	0	0	12	4.8
26.	Culex (Culex) triataaniorhynicus Giles, 1901.	11	4.4	31	12.4	3	1.2	45	18
27.	Culex (Culex) malhari sp. nov.	25	10	52	20.8	39	15.6	116	46.4
28.	Culex (Culex) malkapuri sp. nov.	45	18	13	5.2	25	10	83	33.2
29.	Culex (Barraudius) satarensis sp. nov.	118	47.2	62	24.8	45	18	225	90
30.	Culex (Barraudius) mirjensis sp. nov.	68	27.2	37	14.8	13	5.2	120	48
	Total	**554**	**221.6**	**329**	**131.6**	**292**	**116.8**	**1177**	**470.8**
31.	Armiger (Armiger) subalbatus Coquillett 1898.	168	67.2	123	49.2	69	27.6	360	144
	TOTAL	**1632**	**652.8**	**942**	**376.8**	**801**	**320.4**	**3362**	**1344.8**

Table 3: Species Diversity and Vector Abundance in Western Maharashtra

Sl.No.	Mosquito Species	No. of Species Found	No. of Mosquito Species Collected	%	No. of Vector Species Found	No. of Vector Mosquito species collected	%
1.	Anopheles	16	1524	45.33	5	845	55.45
2.	Aedes	5	301	8.95	3	270	89.70
3.	Culex	9	1177	35.01	4	621	52.76
4.	Armiger	1	360	10.71	0	0	0.00
	Total	**31**	**3362**		**12**	**1736**	**51.64**

contributed 41.7 per cent. Five vector species among the 16 *Anopheles* species (55 per cent), three vector species (93.02 per cent) among the 5 *Aedes* species and four vector species (52.7 per cent) among the 9 *Culex* species were found in Western Maharashtra. The total 12 vector species were contributed 51.93 per cent population. Results indicate that the vector species were prominent and this abundance was alarming sign for the mosquito borne diseases in Western Maharashtra (Table 3).

The seasonal prevalence of mosquitoes in Western Maharashtra reveals that the densities of *An. stephensi* were maximum during the pre monsoon period (Feb – May) (Table 2). The density of *An. subpictus, An. annularis, Culex vishnui, Culex tritaeniorhynchus, Culex bitaeniorhynchus* were prominent during the post monsoon period (Oct-Jan). The density of *An. culicifacies, An. fluviatilis, Aedes albopictus, Aedes aegypti, Aedes vittatus, Culex quinquefasciatus* and singular *Armiger subalbatus* were found abundant in monsoon period (Jun–Sept).

Among newly described species *Anopheles kolhapuri, Anopheles karveeri, Anopheles krishnai, Aedes rajashri, Culex malkapuri, Culex satatarensis* and *Culex mirjensis* were found abundant in monsoon period. *Anopheles mahabaleshwari, Anopheles waii, Aedes sathei* and *Culex malhari* were found abundant in post monsoon period.

Discussion

Mosquitoes respond to temperature increase in various ways. Within limits higher temperature was means more rapid development of larval populations and for shorter time between the blood meals, quicker incubation time for pathogen infection and shorter life span of adults although the latter is dependent on the humidity (Roussel, 1998). The temperature of breeding places plays an important role in the persistence and growth of larvae. Rate of development of larvae accelerates in the warm water and slows down in the cold water (Ramchandra Rao, 1984). Many species such as *An. culicifacies, An. subpictus, An. vagus* breed and survive both in open water such as burrow pits or river bed pools and also in some shady places while *An. fluviatilis* and *An. minimus* prefer to breed in shady places such as under overhanging trees and bushesh or from thick growth of grass. Species breed in deep wells such as *An. stephensi* and *An. varuna* remains in shade most of the day. In fact, *An. stephensi* grows well in cistern or covered wells, which never get any direct sunlight (Ramchandra Rao, 1984).

Vector control requires through knowledge on the ecology of the local species with respect to breeding and resting habitats and behavior. Therefore, periodic survey of vector populations in a given area is most essential for better understanding of the changing ecology, bionomics of mosquitoes and thereby possible disease outbreak prediction and for effective vector control initiation. Keeping in view all above facts the present work was carried out. The rate of mosquito born diseases again depends on the index of species of the region. The rapid separation and identification of mosquitoes of primary medical importance is an important task in the assessment of disease potential area.

Senior White (1937) reported twenty three anopheline species in 1935-36 on Jaypore Hills in which *Anopheles fluviatilis, An. varuna* had the highest peak in the month of Sept. and Feb. These species were most prevalent in the spring and in the rains. *An. jeyporiensis* and *An. culicifacies* were found resting mostly in cattle shade. Senior White *et al.* (1943) again reported eighteen anopheline species from Orissa during the year 1935–1941. *Anopheles annularis* and *An. aconitus* were most prevalent species observed in November-March.

According to Foot and Cook (1959) mosquitoes have thirty two regions in the world. In India and Sri Lanka *Anopheles culicifacies* is a most important and wide spread vector of malaria. This species was also most important vector in Sri Lanka. In the foot hill areas of peninsular India, *Anopheles fluviatilis* was dominant vector species for causing malaria. Foot and Cook (1959) also visualized *Anopheles varuna* from East-central India. *Anopheles stephensi* from Northern West-coast and Gangas plain and *Anopheles sundaicus* from Kolkata while, from Orissa they documented *Anopheles annularis. Culex annulifera was* the chief vector of filarial in the lower parts of the Gangas River basin, Bihar and Orissa on the North-east coast and Travancore state and *Aedes aegypti* was common throughout India (Foot and Cook, 1959). They stated that, *Aedes albopictus* was not so closer to man and less universally distributed in India.

Seasonal abundance of 11 species of *Anopheles* mosquitoes have been reported by Sen *et al.* (1960) from Dhanbad area from 1953 to 1958. *Anopheles culicifacies* was most abundant during monsoon from July to September and also in the month of February. *Anopheles subpictus* was the most predominant species and was found throughout the year with definite seasonal abundance during June to September. *A. annularis* was found during winter. *Anopheles pallidus* was found throughout the year but the peak densities were during winter (November).

Nagpal and Sharma (1983) reported seasonal abundance of twenty four species belonging to five genera of mosquitoes in the South, middle and North Andaman Island during a study tour in January-February 1982. In sixteen species of *Anopheles* the most prevalent species was *Anopheles vagus* followed by *Anopheles kochi* and *Anopheles sunduicus* and most dominated during monsoon. Among Culicine species the most dominant genera was *Culex* and most prevalent species was *Culex quinquefasciatus* followed by *Culex tritaeniorhynchus* and *Culex vishnui* and dominant during the summer and fall in the winter. The genus *Aedes, Armigers* and *Mansonia* reported extremely low in the study area.

Similarly *Anopheles vagus* and *Anopheles subpictus* exhibited a peak during the months of monsoon rains, while *An. culicifacies* a rural vector showed two peaks of abundance, one during the monsoon and other in February (Kaul *et al.*, 1982). Nagpal (1983) also studied seasonal abundance of twenty nine mosquitoes from Nainital Terai (U.P.) belonging to 8 genera *viz. Anopheles* (18), *Aedeomyia* (1), *Aedes* (2), *Armigers* (1), *Coquillettidea* (1), *Culex* (4), *Mansonia* (1) and *Mimomyia* (1) during 1980-82. Survey revealed that *Anopheles subpictus* dominant species during Sept. 1980 and Sept-Oct. 1981, then followed by *Anopheles culicifacies* and *Anopheles fluviatilis*. However, during May-June 1981 survey *Anopheles culicifacies* was most prevalent species followed by *Anopheles subpictus* and *An. annularis*. During the Jan. – Feb. 1982 survey, *Anopheles fluviatilis* was most prevalent species followed by *Anopheles splendidus* and *Anopheles culicifacies*. *Culex quinquefaciatus* was the most prevalent species among the Culicinae collections in all four surveys followed by *Culex tritaeniorhynchus* and

Culex vishnui. The seasonal abundance of *Anopheles culicifacies* and *Anopheles fluviatilis* were distributed throughout the Ferial belt, while *Anopheles culicifacies* exhibited two seasonal peaks *i.e.* in May-June and Sept.–Oct.

Das *et al.* (1984) studied seasonal abundance of forty two species belonging to a genera *viz.*, *Anopheles, Culex, Aedes, Mansonia, Armigeres* and *Coquillettidia* of mosquitoes from various places of Meghalaya during April-May 1980. Out of which *Anopheles vagus, Anopheles annularis, Anopheles barbirostris, Anopheles philippinensis, Culex tritaeniorhynchus, Culex vishnui, Culex bitaeniorhynchus, Culex gelidus* were most prevalent species throughout year and *Mansonia indiana, Mansonia uniformis* and *Aedes albopictus* were extremely rare.

Fifty one species of *Anopheles* and their locality, taxonomy, seasonal abundance, distribution, adult bionomics, larval ecology and diseases have been attempted by Rao, (1984). Nagpal and Sharma (1987) documented seasonal abundance of sixty one species from Assam, Meghalaya, Arunachal Pradesh and Mizoram during Sept. 1986. These sixty one species belongs to eight genera *viz. Anopheles, Aedes, Armigeres, Coquiletidia, Culex, Malaya, Mansonia* and *Toxorhynchites.* The most dominant genus was *Anopheles* then by *Culex, Aedes* and *Mansonia.* The most prevalent species in the genus *Anopheles* was *Anopheles vagus* followed by *Anopheles nigerrimum* and *Anopheles nivipes.* In the genus *Culex* the most prevalent species was *Culex quinquefasciatus* followed by *Culex tritaeniorhynchus* and *Culex vishnui. Aedes albopictus* was the most dominant species in the genus *Aedes* and it was followed by *Aedes chryolineatus* and *Aedes aegypti.* In the genus *Mansonia* the

200 | Mosquito Diversity and Control

most common species was *Mansonia annulifera* followed by *Mansonia uniformis* and *Mansonia indiana*. The other four genera *viz. Armigeres, Coquillettidia, Malaya* and *Toxorhynchites* were rare in the region.

Proliferation of mosquitoes is determined by the availability of suitable and sufficient habitat for the larval stages, resting and feeding sources nearby. In urban areas like Delhi the *Aedes aegypti* populations increased with the onset of monsoon rainfall in June – July (Reuben *et al.*, 1973). Prolific breeding could be seen in manmade habitats including air coolers. Ruben *et al.* (1992) visualized ten species of *Culex* mosquitoes in Madurai, Southern India. *Culex tritaeniorhynchus, Culex pseudovishnui* and *Culex vishnui* were feed dominantly on cattle, but less frequently on humans and on pigs and birds. These three species were occurring predominantly throughout year where the cattle were reared, but *Culex tritaeniorhynchus* and *Culex vishnui* showed a marked in the population during the hot season. *An. culicifacies* was recorded in high numbers from April to September while *An. fluviatilis* only during October to March in Uttaranchal (Shukla *et al.*, 2007).

From Maharashtra recently, Sathe and Girhe (2001) reported four species of *Anopheles* namely, *Anopheles culcifacies, Anopheles stephensi, Anopheles theobaldi* and *Anopheles subpictus*. The most prevalent species of *Anopheles* in Kolhapur region was *Anopheles culicifacies* while, *Anopheles subpictus* was rare in the Kolhapur region.

Sathe and Girhe (2002) reported fifteen species of mosquitoes from Kolhapur district belonging to genera *Anopheles* (4) *Culex* (3) and *Aedes* (7). Out of which *A. culicifacies, C. pipiens* and *A. aegypti* were predominant throughout the year. While, *A. indica* S. and G. the largest mosquito species found was extremely rare. Other eleven species were moderately distributed in Kolhapur.

Girhe and Sathe (2001) studied incidence of malaria during the year 1992-1996, was increasing in order. Maximum, 700 infection cases were reported during the year, 1996 due to prevalence of *Anopheles* mosquitoes. Later, incidence of malaria declined from the years 1997-2000 from Kolhapur region.

Recent work of Sathe and Girhe (2001, 2002) refer to the following species of genus *Aedes* namely, *Aedes aegypti, Aedes kolhapurensis, Aedes indica, Aedes indicus, Aedes sangiti, Aedes panchgangi* and *Aedes uniformis* from Maharashtra. The work of Sathe and Girhe (2002) reported the first record of largest species *Aedes indica* from the world. *Aedes aegypti, Aedes indicus, Aedes uniformis* are well known to science from Southern Maharashtra. Four species of *Culex* namely *Culex epidesmus, Culex pipiens, Culex modestus, Culex malayi* previously have been reported from Maharashtra (Sathe and Girhe, 2002).

Murty *et al.* (2002a) studied the seasonal prevalence of *Culex quinquefasciatus* in the rural and urban areas of the East and West Godavari districts of Andhra Pradesh, India during 1999. These species occur dominantly throughout year in rural and urban areas. Murty *et al.* (2002b) reported a seasonal abundance of *Culex vishnui* sub group and *Anopheles* species in an endemic district of Andhra Pradesh during 1999. *Culex vishnui* subgroup was dominant throughout the year. Their density was high in January-December 1999. Kanojia *et al.* (2003) reported seasonal abundance of mosquitoes in Gorakhpur district, Uttar

Pradesh during 1990 to 1996. The seasonal fluctuations in the mosquito population was recorded. High prevalence of *Culex quinquefasciatus* was observed in March, *Culex tritaeniorhynchus* predominant species was noticed in September. The other species such as *Culex pseudovishnui*, *Culex whitmorei*, *Culex gelidus* and *Mansonia uniformis* had also peak occurrence in September. Seasonal prevalence of *Anopheles* species *Anopheles subpictus* and *Anopheles peditaeniatus* showed high prevalence during July and September.

Sharma *et al.* (2005) studied seasonal prevalence of *Aedes aegypti* in Delhi during 2003. *Aedes aegypti* was abundant in month of August and September. The rise in breeding indices during the post monsoon season may be attributed to increase in artificially collected breeding containers due to rains. Joshi *et al.* (2005) give seasonal prevalence of Anopheline mosquito in irrigated and non-irrigated area of Thar, Rajasthan during August 2001 to July 2002. They reported Anopheline species, *Anopheles subpictus*, *Anopheles culicifacies* and *Anopheles stephensi*. During monsoon *Anopheles subpictus* and *Anopheles stephensi* were dominant species but was peak in August to October. While *Anopheles subpictus*, *Anopheles stephensi*, *Anopheles culicifacies* and *Anopheles annularis* reported in winter season in irrigated area and *Anopheles culicifacies* and *Anopheles annularis* were predominant in summer during April to July.

Tilak *et al.* (2006) studied the seasonal prevalence of mosquitoes in Pune during 2001 to 2003. They reported seventeen species of five genera *Anopheles*, *Culex*, *Aedes*, *Armigeres* and *Mansonia*. The dominated genera were *Culex* followed by *Anopheles*, *Aedes Armigeres* and *Mansonia*. The seasonal abundance of mosquito in Pune reveals that the densities of *Anopheles stephensi, Anopheles varuna* and *Anopheles vagus* were maximum during summer (March – May). The density of *Anopheles annularis* and *Anopheles stephensi* were high in rainy season (June-Sept.) and winter season (Nov.–Feb.). The Culicines *Culex quinquefaciatus* was found in higher densities in all the three season with abundance in rainy season. Whereas the abundance of the other Culicines *i.e. Culex cornutus, Culex gelidus, Culex sitiens* and *Culex univittatus* were higher in summer season. *Aedes aegypti* was found in all the three season with high prevalence in rainy season. *Armigeres* and *Mansonia* species were extremely rare and reported in winter season.

Pemola and Jauhari (2006) reported ten *Anopheles* and *Culicine* mosquitoes in the Doon valley Dehradun, Uttaranchal during 1999-2002. The seasonal prevalence of Anopheline species *i.e. Anopheles culicifacies, Anopheles fluviatillis* and *Anopheles stephensi* were dominant in monsoon (June-Sept.) and post-monsoon (Nov.-Dec.). The seasonal prevalence of Culicine species *Culex mimeticus, Culex vishnui, Culex quinquefasciatus* and *Aedes albopictus* were dominant in between May to November and December to February. Malarial Research Centre (MRC) (2006) reported seasonal prevalence and bionomics of *Anopheles culicifacies, Anopheles fluviatilis, Anopheles minimus, Anopheles sundaicus* and *Anopheles stephensi* in Delhi, Kheda (Gujarat), Bhaber (Uttar Pradesh) and Rourela (Orissa) during 1989-1991. In Delhi seasonal prevalence were studied in a riverine zone of the river Yamuna and in a non riverine belt. *Anopheles culicifacies* was most dominant species in the riverine zone observed in April and in October. The non riverine area, the peak abundance was observed in May and August. *Anopheles culicifacies* was dominant in the

Northern part of the reservoir zone where water pollution was at minimum level. In Khed district (Gujarat) *Anopheles culicifacies* was found throughout the year in varying proportions. In the canal-irrigated area, its density starts to build up from February and reach high in March. In the non canal-irrigated areas, the abundance of *Anopheles culicifacies* remains low throughout the year. In Bhabar area of Uttaranchal in north India, *Anopheles culicifacies* abundance remains low during January to June and October to December. It increases during monsoon reaching a high in August.

Baruah *et al.* (2007) reported seasonality of malaria in Lama Camp, Hoograjuli, Behali and Pabhoi area in Sonitpur district, Assam during 2002 to 2003. The study period was grouped into four seasons such as pre-monsoon (March to May), Monsoon (June to August), post-monsoon (September to November) and winter season (December to February). They reported seven species of genus *Anopheles i.e. Anopheles annularis, Anopheles culicifacies, Anopheles dirus, Anopheles fluviatilis, Anopheles minimus, Anopheles philippinensis* and *Anopheles varuna.* The *Anopheles philippinensis* dominated in all the four study area followed by *Anopheles annularis, Anopheles minimum, Anopheles culicifacies, Anopheles fluviatilis, Anopheles dirus* and *Anopheles varuna.* Density of *Anopheles philippinensis, Anopheles annularis, Anopheles minimus, Anopheles culicifacies* and *Anopheles dirus* increased during the pre-monsoon period, peak in monsoon and declined during the post-monsoon.

From Southern Maharashtra, Tingare and Sathe (2007) reported *Aedes khanapuri* sp. nov., *Aedes rhadhanagari* sp. nov., *Aedes tasgaonensis* sp. nov. and *Aedes mangalvedhi* sp. nov. are newly described for the first time from India. From the genus *Anopheles*, six new species have been described for the first time *i.e.* *Anopheles atpadi* sp. nov., *Anopheles akuluji* sp. nov., *Anopheles sagareshwari* sp. nov., *Anopheles karmalae* sp. nov., *Anopheles ajrae* sp. nov. and *Anopheles mirajensis* sp. nov. Seven species from genus *Culex* were newly described and reported for the first time from Southern Maharashtra, which includes *Culex solapurensis* sp. nov., *Culex krishnai* sp. nov., *Culex chandrabhagi* sp. nov., *Culex rankali* sp. nov., *Culex mahalaxmi* sp. nov., *Culex kalambae* sp. nov. and *Culex sangolensis* sp. nov.

Recently, Jagtap and Sathe (2008a) studied three decades trend of malaria situation of Sangli district during the period 1971-2005. The maximum amplification of the disease was observed in drought prone area *i.e.* Jath, Kavathemahankal and Atpadi. This is correlated with rainfall. In early epidemic phase *Plasmodium vivax* was dominant but, recently increase in more than 30 per cent trend of *Plasmodium falciparum* was observed. Jagtap and Sathe (2008b) documented the role of intensified mass surveillance campaign in malaria problematic section in Sangli district. Jagtap and Sathe (2008c) studied the Chloroquine resistance to *Plasmodium falciparum* species in Etapalli block of district Gadchiroli.

Very recently, Sathe and Jagtap (2009) studied the tree hole breeding and resting of mosquitoes in Western Ghats. Total 106 tree holes were examined, 32 tree holes were found positives for adults and six were found for larvae. Jagtap and Sathe (2009) studied the incidence of dengue and shifting trend to rural in Kolhapur district. More recently, Sathe and Jagtap (2010) studied the abundance of Anopheles mosquitoes from Western Ghats and found responsible for malarial incidence in the region.

Chapter 6
Life Cycle of Mosquitoes

There are over 3,500 different species of mosquitoes throughout the world. There are four stages of life cycle of a mosquito: egg, larva, pupa and adult. To develop the eggs female mosquito needs the blood meal. Male mosquitoes do not bite - they feed on plant nectar.

Eggs

Female mosquitoes can develop 200 to 300 hundred eggs at each blood meal and lay them in water. A female anopheline mosquito normally mates only once in her lifetime. She usually requires a blood meal after mating for development of eggs. Blood meals are generally taken every 2-3 days before the next batch of eggs is laid. Oviposition sites vary from small hoof prints and rain pools to streams, swamps, canals, rivers, ponds, lakes and rice fields.

Larva

A larva hatches from the egg after about 1-2 days called

"wrigglers". There are four larval stages or 'instars'. The total time spent in the larval stage is generally 8-10 days at normal tropical water temperatures. At lower temperatures, the aquatic stages take longer to develop. These are found in all kinds of standing water, such as; ditches, woodland pools and anything that holds water for more than a week.

Pupa

In about 7-10 days after the eggs hatch, larva develop into pupa before becoming adult mosquito. The pupa is comma shaped and is a non feeding stage. It stays under the surface and swims down when disturbed. The pupal stage lasts for two to three days after which the skin of the pupa splits. Then the adult mosquito emerges and rests temporarily on the water surface until it is able to fly.

Adult

Mating takes place soon after the adult emergence from the pupa. The first batch of eggs develops after one or two blood meals, while successive batches usually require only one blood meal. Some mosquitoes enter houses to bite and are described as being endophagic; others bite mostly outside and are called exophagic. Host preferences are different for different species of mosquitoes. Some mosquitoes prefer to take blood from humans rather than animals and are described as being anthropophagic/anthropophilic while others only take animal blood and are known as zoophagic/ zoophilic. The feeding habits of mosquitoes are unique for man and other animals. The male mosquitoes feed only on plant juices. Female mosquitoes feed on man, domesticated animals, birds, wild animals, snakes, lizards, frogs, and toads. If mosquitoes do not get blood meal, they will die without laying viable eggs.

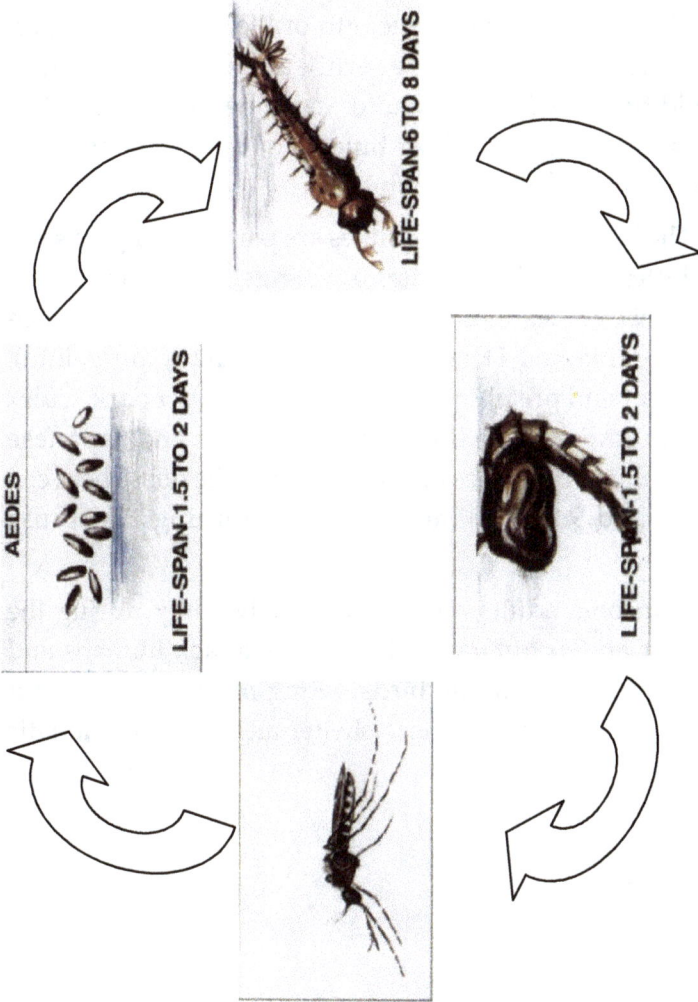

Plate 28–Figure 176: Stages of Life Cycle

The flight range of mosquitoes depends again on the species. The flight range of females is longer than males. Many times wind is a factor for migration of mosquitoes. Some mosquitoes have been recorded as far as 12 km from their breeding source. The length of life cycle of the adult mosquito depends on several factors: temperature, humidity, sex of the mosquito and time of year. Most of males live for a very short time, about 8 to 10 days; and females live for about a month.

The bite of Aedes mosquitoes are painful and persistent day bitter. They do not enter dwellings, and prefer to bite mammals. Aedes mosquitoes are strong fliers and causes Chikungunia and Dengue Fever. Culex bites are painful, persistent and prefer to attack at dusk and after dark. Culex are generally weak fliers causes Filaria and Japanese encephalitis. Anopheles mosquitoes enter the house between 5 p.m. and 9.30 p.m. and in early morning and cause Malaria.

Mansonia adults appear to be active only during the summer and autumn months; readily attack humans and other animals including birds, biting mostly at night but also during the day in/near shelter and cause Lymphatic Filariasis.

Chapter 7
Mosquito Control

Introduction

Malaria, filaria, dengue, chikungunya and Japanese encephalitis are major vector borne diseases of India. The mosquitoes are most important among several insect and acarine vectors of human diseases. Because they transmits diseases like malaria, filariasis, Japanese encephalitis, chikungunya and dengue. The diseases such as plague (flea borne), scrub typhus (mite borne), and KFD (tick borne), are of occasional and sporadic occurrences and have lesser importance from the viewpoint of their routine vector control. Therefore, only the conventional method of mosquito control refers to use of chemical insecticides and biocontrol agents. Mosquito can complete its life cycle from egg to adult within 6-10 days. Hence they buildup their population in short time and becomes difficult to control. Untreated standing waters, floods or irrigations are the causes for epidemics. Such situations require special efforts for the control of vectors where as chemical insecticides are required as a short-term measure to control epidemic.

The mosquitoes can be controlled not only by chemicals but also by physical and biological methods. Moreover biological methods are ecofriendly. Therefore, emphasis should be given on ecofriendly measures. Vector management is disease management. Therefore, through the best and latest techniques and depending upon local situation, control of vectors should be made with minimal adverse side effects on humans, animals and the environment.

Industrial development, urbanization, agriculture, etc, have increased the population of mosquitoes much beyond their natural levels. Mosquitoes transmit diseases from man to man, and animals to man, and also cause serious annoyance problem. Therefore, for avoiding nuisance, biting, and preventing the spread of mosquito borne diseases, their control is must. The people should realize the magnitude of the health threat by the mosquitoes. Some of the world's most dreaded diseases are known to be transmitted by mosquitoes. India is facing the problem of malaria, filariasis, Japanese encephalitis, chikungunya and dengue every year. Therefore, as an integral part of the disease control strategy, vector control is must.

Displaced populations have an increased risk of vector-borne diseases. Factors that make displaced populations more susceptible to vector-borne diseases include the following:

Stress, lack of good nutrition, and lack of previous exposure to the disease will lower a population's immunity to vector-borne diseases. This is especially true with malaria. In urban or highland areas there may be very little exposure to malaria; In warmer climates, there is an increased chance for the disease to be transmitted. Displaced populations can transfer certain parasites and diseases from their former homes to new locations where they multiply and spread and becomes susceptible.

Displaced populations are more exposed to vectors due to:

1. Overcrowding (spread from person to person)
2. Poor housing (results in closer contact with sandfly vectors of leishmaniasis, flea vectors of rodentborne diseases, or tick-borne relapsing fever).
3. Increased number of breeding sites leads mosquito multiplication

Mosquito populations can multiply in great numbers in poorly drained water distribution points, either due to more pools of water or more domestic water containers. But, this can increase the incidence of mosquito-borne diseases:

☆ More water-storage containers increase breeding of the dengue fever vector _Aedes aegypti_.

☆ More water-filled pit latrines increase breeding of the encephalitis vector _Culex quinquefasciatus_.

☆ More groundwater pits, ponds, and footprints increase breeding of the malaria vector _Anopheles gambiae_.

While, evacuation sites and newly established camps may have severe problems with flies, lice, and mosquitoes, to lead several diseases. Natural disasters (_e.g._, El-Nino floods and hurricanes in 1997-98) may change the environment and increase the breeding sites of vectors and diseases.

Man's effort to control mosquitoes is an on-going battle since eradication of mosquitoes is impossible. The benefits of mosquito control are to reduce the threat from mosquito-borne diseases, to enjoy quality life. However, the risk associated with mosquito control by chemicals is also of major concern to human health and the health of our environment. Integrated Vector Management is the basic approach and ecofriendly. Adoption of more than one method of control which can be utilized in a cost effective manner is integrated vector management.

Source reduction of a vector and improved sanitation play important role in vector control. Other control measures can be used as supplement.

Biological control of vectors, through use of predators and pathogens, particularly larvivorous fish *Gambusia affinis* and *Poecilia reticulata* (Guppy) has been a useful eco-friendly vector control approach. Larvivorous fishes can be used in ponds, unused wells etc.

Insecticides are to be used judiciously and selectively in carefully identified areas which depends on the epidemiological and entomological parameters. Chemical methods can supplement to basic sanitation and environmental management even when they are employed as principal means to achieve rapid and maximum control of vector. Vector management is possible by:

1. Selective indoor residual insecticide spray in high risk areas.

2. Promotion of use of nets preferably Insecticide Treated Nets (ITN).

3. Selective chemical control of vector larvae/pupae (aquatic developmental stages).

4. Space spray/aerosol application/fogging.

The goal of mosquito control should be "To control mosquitoes in a safe, efficient, and cost-effective manner, and while doing so prevent damage to humans, animals, and the natural environment". This task can be successfully accomplished through integrated control approach. IMM plan should be based on sound scientific knowledge on bionomics, ecology, and use of the latest technology about equipment and materials. These methods would furnish a cost-effective level of mosquito suppression to protect man and domestic animals from vector bites, and the diseases.

Mosquito Control Strategies

1. Identification of vectors of human diseases?

2. Identification of mosquito species as the primary source of annoyance?

3. Identification of important breeding sites of different mosquito species?

4. Identification of seasonal pattern of mosquito breeding?

5. Identification of resting places of adult mosquitoes?

6. Identification of feeding preferences of vector mosquitoes?

The above information is very helpful for formulating strategies for appropriate mosquito control measures, which are ecofriendly and less costly. Modern integrated mosquito management programme requires a vast array of knowledge and competence in a number of technological specialties.

Vector (mosquito) control has basic two components:

1. Preventive control and
2. Curative control

1. Preventive Control

"Prevention is better than cure" is application in pests, diseases and vectors. For mosquitoes following preventive control measures are adopted.

(*i*) Destruction of Breeding Sites/Places of Mosquitoes

The very sure method of mosquito control is destruction/ elimination of breeding sites/places of mosquitoes. Mosquito breeds in water/aquatic medium. Therefore, destruction/ elimination of their breeding sites play a very crucial role. Drainage system should not be open type it should be closed type. Draining water supply should not have any leakage, that water becomes good source for mosquito breeding. Drinking and other usable water should not be kept open; it should be kept in closed condition for avoiding breeding of mosquitoes. Unused vehicle tyres are good source for breeding mosquitoes; tyres should not kept open and should not collect rainwater or any other type of water.

(*ii*) Environmental Management

This involves any change that prevent mosquito breeding and their prevalence.

(*iii*) Environmental Modification

Making environment unfavorable to mosquito breeding and survival; storage of water, storage of water supply, keeping water containers/tanks closed and several other methods.

(iv) Environmental Manipulation

The main objective of this method is to deprive the proper environment of mosquitoes so that the mosquito will not survive, will not breed and multiply.

(v) Personal Protection

Use of protective clothing, mats, coils, aerosols.

(vi) Use of Repellants

DEET, Citronella oil, neem oil, essential oils etc. are used as repellants.

2. Curative Control

There are several strategies of curative control of mosquitoes. Important curative control measures are given below:

(i) Chemical Control

Chemical control has two components *i.e.* larvicidal control and adulticidal control.

(a) Larvicidal Control

Larval surveys: The larval surveys, can be done with basic sampling unit the house or premise, for water holding containers. Containers are examined for the presence of mosquito larvae and pupae. Depending on the objective of the survey, the search may be for *Aedes* larvae or *Anopheles*. Four indices are commonly used to monitor *Aedes aegypti* infection levels:

1. *House index (HI)*: percentage of houses infected with larvae and/or pupae.

$$HI = \frac{\text{Number of Houses Infected}}{\text{Number of Houses Inspected}} \times 100$$

2. *Container index (CI)*: Percentage of water holding containers infected with larvae or pupae

$$CI = \frac{\text{Number of Positive Containers}}{\substack{\text{Number of Containers} \\ \text{Inspected}}} \times 100$$

3. *Breteau index (BI)*: Number of positive containers per 100 houses inspected

$$BI = \frac{\text{Number of Positive Containers}}{\text{Number of Houses Inspected}} \times 100$$

4. *Pupae index (PI)*: Number of pupae per 100 houses

$$PI = \frac{\text{Number of Pupae}}{\text{Number of Houses Inspected}} \times 100$$

Anopheles larvae found in fresh water. Therefore, after proper survey of breeding sites and situation control measures are started. Paris green, temiphos, fenithion, pirimiphos methyl, methoprene (growth regulator), biocides and neem acts as mosquito larvicides and are widely used in mosquito larvae control programmes.

1. Temiphos (50 per cent EC)

Temiphos (Abate) is one of the organophosphorous insecticide used for mosquito control. Chemically it is 0,0,0,0 tetra methyl 0,0 thio diphenylephosphorothioate. It has very low toxicity to other aquatic animals, birds, insects and humans and highly effective against mosquito larvae. This contact larvicide is recommended in potable and clean water.

2. Fenithion (82.5 per cent w/k)

Fenithion has long residual effect as larvicide on mosquitoes. It is organophosphorous compound of quick killing effect. The compound has relatively high toxicity and mainly applicable in polluted water. 5 ml fenithion is mixed in 10 liters and sprayed by compression sprayer (175-225 liters/hectare) or 100 ml of formulation (82.5 per cent w/ v)/hectare/week. This is also used as 1 liter for 50 linear meters of water surface. Hand compression/Knapsack sprayers are ideal for application. The dose is applicable once in a week by exempting potable water.

3. Pirimiphos Methyl (50 per cent EC)

Pirimiphos Methyl is also from organophosphorous group of insecticides. It is soluble in most organic solvents and effective against mosquito larvae. This larvicide is not recommended in clean water but, used in polluted water. 20 ml Pirimiphos Methyl mixed in 10 liters and sprayed by compression sprayer (175-225 lit./hectare) or 400 ml of formulation (50 per cent EC)/hectare/week. This is also used as 1 lit. for 50 linear meters of water surface. Ideal sprayers for this compound are Hand compression/ Knapsack sprayers. The dose is applicable once in a week.

The another larvicide is prepared biologically by bacterium, *Bacillus thuringiensis israelensis* (Bti) H- 14 is very usefull in killing mosquito larvae.

(b) Adulticidal Control

Indoor Residual Spraying

House spraying pesticides remains a valuable tool in malaria control when applied in the right circumstances. Indoor residual spraying with DDT was the major success of malaria control in the 1950s and 1960s. Malaria was eradicated, or almost eradicated, from many parts of the world during this time. However, continued application of insecticides is not sustainable as vector may develop resistance to insecticides.

Indoor Residual Spraying should be Based on

1. A high percentage of the structures in an operational areas have adequate sprayable surfaces and can be expected to be well sprayed;
2. The majority of the vector population is endophilic, *i.e.* rests indoors;
3. The vector is susceptible to the insecticide in use.

The efficacy and persistence of residual vary with the type of surface sprayed *i.e.* mud, wattle, wood, thatch, palm leaf or asbestos. Hence spraying should be done according to specified criteria.

Criteria for Selective Application

1. Considerable resources required for indoor residual spraying, combine insecticides for avoiding resistance and possible environmental hazards.
2. Pesticides are recommended only in high priority areas.
3. Frequencies and times of applications must be decided.
4. The spraying programme must justify cost-effectiveness.
5. Epidemiological impact should be assessed by correlating the quality and coverage of spraying operations with malaria trends. This is because the transmission and the burden of malaria are often

focal and may vary with malaria endemicity and vector density/population.

6. Operational boundaries must be more precisely defined.

7. Spraying operations may be limited to certain geographical areas, individual villages or groups of villages, or specific times of the year.

8. Selective insecticides reduce costs and selection pressure for resistance, hence recommended.

9. Reorientation of large-scale spraying programmes be adopted.

Selection of Insecticides

The following factors are taken into account for selection of an insecticide for indoor residual spraying.

☆ *Residual effectiveness*: Maximal residual effectiveness is desired with safety of man and environment.

☆ *Vector susceptibility* : Vector should be susceptible to insecticide.

☆ *Impact on disease*: The insecticide should reduce the incidence of disease.

☆ *Excito-repellency*: The exciter-epellent effects of insecticides are to be fully understood and must be taken into account in operational activities.

☆ *Indoor Residual spray (IRS)*: Most of the insecticides having residual effect are sprayed indoors, so that mosquitoes after having bite on an infective person will rest in the house and will pick up sufficient insecticide particles sprayed on the walls and other indoor surfaces of the house and will die.

☆ *Target area*: All the interior walls and ceilings are treated. Permanent human dwellings, field huts where people sleep during the planting or harvesting season should be sprayed.

☆ *Selection of Insecticides*: The selection of an insecticide spraying should be based on availability, cost, residual effectiveness, safety, vector susceptibility and excito-repellency. There are large number of insecticides, which are used as adulticides for indoor residual spray. These are DDT, Malathion and different formulations of synthetic pyrethroids.

Insecticide Formulations Used Under NVBDCP

The following formulations/compounds are used under the NVBDCP for control of malaria:

(i) DDT (Dichloro-diphenyl-trichloroethane)

DDT is invented by Paul Muller in 1946. It is used for malaria control since 1946 in India. Recently, there has been a tendency to curb the use of DDT due to its persistence in the environment. DDT is applied in agriculture, it contaminate water resources, enters the biochain and at each step of the biochain, it gets more concentrated (bio-magnification) till it reaches human beings. It is stored in the human body fat and released in the human milk. Ultimately reaches the infants right from the time of birth DDT is persists however, in spite of its extensive use in agriculture, no adverse reaction of DDT on human health has been reported so far. Therefore, WHO has recommended that at this stage there is no justification on toxicological or epidemiological grounds for changing current policy towards indoor spraying of DDT for vector-borne disease control. DDT is therefore used for vector control, since,

1. It is used only for indoor spraying
2. It is effective
3. The material is manufactured to the specifications issued by WHO
4. The necessary safety precautions are taken in its use and disposal.

Govt. of India has constituted a mandate Committee on DDT which reviews the use of DDT in public health and decides its quantity to be released for the vector borne diseases control programme every year. DDT is cheaper than the other insecticides and even in those areas where resistance to DDT has been recoded in studies with WHO test kits, it showed its excito-repellent action. Hence, useful and used against mosquitoes. DDT is required 150 MT per million population for two rounds of spray. In areas where third round is proposed in selected villages, additional requirement of 75 MT per million population should be estimated.

(ii) Organophosphorous Compounds

Malathion 25 per cent WP is used under the progamme in areas with DDT resistance. The disadvantage of organophosphorous compounds is that although they are used in agriculture for crop protection only once or twice a year, the spray squads of residual insecticide in the human dwellings works for 6 or 7 months. This long exposure results in acute toxic symptoms and if not controlled properly may lead to mortality. Therefore, sprayers of OP should be provided more elaborate protective garments and their blood cholinesterase level is to be checked periodically to assess the toxic impact upon them. These compounds are also toxic to domestic animals.

Under Indian conditions, three rounds of spray with organophosphorous compounds are given as against two rounds of spray with DDT.

Special Instruction

In case of OP poisoning, the patient should be transported as soon as possible to a doctor to receive an antidote. In organophosphate poisoning, 2-4 mg of atropine should be given intravenously (for children 0.5 to 2 mg according to weight). Depending on symptoms, further doses of 2 mg should be given every 15 minutes for 2-12 hours in severe cases. Automatic injections are available for administration of atropine.

900 MT Malathion 25 per cent per million population for three rounds of spray are required.

(iii) Synthetic Pyrethoids

In India pyrethroids are new insecticides introduced for control of vector borne diseases. Cost is much higher than the cost of DDT and Malathion.

(*a*) Deltamethrin 2.5 per cent WP, (*b*) Cyfluthrin 10 per cent WP, (*c*) Alphacypermethrin 5 per cent WP (*d*) Lambdacyhalothrin 10 per cent WP and (*e*) Bifenthrin 10 WP are recently restricted for controlling mosquitoes in India by Insecticide Board, India.

Special Instruction

In treating pyrethroid poisoning vitamin E oil preparations can be given for prolonged paraesthesia. Only in cases of definite allergic symptoms should corticosteroids be administered. On occurrence of convulsions after sever intoxication, intravenous injection of 5-10 mg Diazepam (or other benzdiazepine derivatives) should be given.

Requirement of Synthetic Pyrethroids

1. Deltamethrin 2.5 per cent WP: 60 MT per million population for two rounds of spray. In some areas, where a further round is required in selected villages, additional requirement of 30 MT per million for the population of selected villages is estimated.

2. Cyfluthrin 10 per cent WP: 18.75 MT per million population for two rounds of spray and 9.38 MT per million population for selected villages as a special round/third round of spray.

3. Lambdacyhalothrin 10 per cent WP: 18.75 MT per million population for two rounds of spray and for a special round or IIIrd round of spray in selected villages, 9.38 MT per million population is recommended.

4. Alphacypermethrin 5 per cent WP 37.5 MT per million population for two rounds of spray is essential.

5. Bifenthrin 10 per cent WP : 18.75 MT per million population for two rounds of spray and 9.38 MT per million population for selected villages as special round/third round of spray is required.

(ii) Environmental Control

Alter breeding sites by draining or filling sites, regular disposal of refuse, maintain clean shelters, and personal hygiene.

(*iii*) Mechanical Control

Use screens or bednets, traps, food covers, lids or polystyrene beads in latrines.

(*iv*) Biological Management

Use living organisms or products against vector larvae, such as fishes that eat larvae (*e.g.*, tilapia, carp, guppies), bacteria (*Bacillus thuringiensis israelensis*) that produce toxins against larvae, free-floating ferns that prevent breeding, etc

Vector Control Strategies

Effective vector control strategies are based on four facts:

1. Knowledge and understanding of vector biology
2. Surveillance of vector species
3. Incrimination of vector species
4. Public education and implementation of effective control measures.

Vector control programme in India, as in the case with many anti-malaria programme elsewhere, in the world, mostly rely on usage of natural and synthetic chemical molecules, which have potential to kill the target insects. Vector control strategies may range from simple treatments (self-protection and home improvement) to more complex measures that require participation from vector control experts (entomologists).

Community Participation

Involvement of various community components counts the success of vector control. Panchayats should be involved in successful indoor residual insecticide spray. Panchayats/village/local bodies/village heads/Block Development Officers/Mahila Mandals, religious groups etc., are to be informed about the spray schedule at least before a fortnight. This advance information must be mopped up by Surveillance Workers/Malaria Inspectors/District Malaria Officer so as to facilitate the villagers to extend full cooperation in getting actual spray inside of human dwelling with the objective of full coverage of targeted population.

Safe Use of Pesticides

There are three components of safety use of pesticides.

(I) Safe Use and Storage of Pesticides

1. Strict procedures must be followed when handling insecticides and other related equipment.

2. Extra precaution should be taken in choosing insecticides and deciding when, how, and for how long to apply them.

3. Pesticides and the spray machines should never be transported in vehicles, which is taking food.

4. Pesticides must be stored in locked and ventilated buildings.

5. Avoid poisoning: Poisoning may be unintentional, children may play with, the novelty of the situation, and the traumatic experience are dangers of poisoning.

(II) Safe Storage and Disposal of Used Insecticide Containers

1. Follow strict guidelines for storage.

2. Dispose every material came in contact with insecticides.

3. Ensure that the displaced community cannot obtain used pesticide containers.

(III) Safety of the Spray Staff

1. The staff should have had prior training on the safe use of pesticides.
2. The staff should have protective clothing (uniforms, gloves, masks etc.).
3. The staff should never smoke, drink, or eat during the job.
4. The staff should have access to good washing facilities after the job is done. In refugee situations all of the above conditions should be maintained.
5. Guidelines on safety training, medical surveillance, and diagnosis and treatment of insecticide poisoning are available from the World Health Organisation (WHO).

Use of Fogging as a Vector Control Measure

Fog is an aerosol spray of droplets with a Volume Median Diameter (VMD) in the range of below 50 microns (mostly 5-15 microns) with reduced visibility. However, fogging is last option in chemical control because of the following limitations:

1. Temporary stay of fog in the environment with no residual effect.
2. Affect adult mosquitoes that come in contact with the fog.
3. Repeated applications needed.
4. High costs.

5. Effect on mosquitoes are highly dependent on climatic factors like wind velocity and its direction, humidity, temperature, etc.

6. Speed of the movement of fogger carrying vehicle or spray men affects vector mortality.

7. Dispersal of fog.

8. Quality of fogging equipment.

Fogging is found useful for control of vectors in specific situations such as outbreak situations of diseases like Dengue/DHF. Good planning and management, training of operators in application technique, and the efficient calibration and use of equipment are crucial factors for successful control of vectors. However, quality of fogging equipment plays important role in effective control of vectors.

Thermal Fogging

Insecticide is vaporized, which condenses to form a fine cloud of droplets on contact with cooler air when it comes out of the machine. This is the principal of thermal fogging. The insecticide is vaporized at a very high temperature inside the machine. Once the fog comes out of the machine, it tends to spread in different directions by mixing with wind. The insecticide of choice for fogging may be Malathion/ Pyrethrum extract because of relative lower mammalian toxicity and being biodegradable. These pesticides do not persist in environment for longer durations. Thermal fogging is psychologically more acceptable as it generates a highly visible fog. The portable thermal fogger and mist blowers are most commonly used equipments for fogging. Vehicle mounted machines have limitation used at communicable roads. It is much more expensive and also less effective than ULV spray.

Advantages

1. Needs lesser active ingredient of insecticide
2. Fog easily visible.

Disadvantages

1. Formulation contains large volume of diluents (organic solvents) which make operation expensive due to high cost of solvent and application.
2. Thick fog causes reduced visibility and traffic hazards.
3. Burning of large volumes of diluents may not be environmental friendly.
4. Very high temperatures of machine operations and use of organic solvents (highly inflammable) poses serious risk of fire hazards.

Spray Timing

Wind speed, humidity, temperature and rainfall are very crucial climatic conditions that influence impact of aerosol application. The most suitable time for fogging is the evening and early mornings when relatively suitable climatic conditions exist and the mosquito vectors are most active.

Most insects are more active in dusk/dawn and thus late evening/early morning fogging is done. Dengue vectors are more active in morning and afternoon and are found mostly indoor. Hence, best results can be had with early morning and late afternoon indoor space spray for adult dengue vectors.

Table 4: Formulations of Fogging for Indoor and Outdoor

Sl.No.	Name of the Insecticide	Commercial Formulation	Preparation of Formulation	Equipment Used	Remarks
1	Pyrethrum extract	2.0 per cent extract	1: 19 *i.e.* 1 part of 2 per cent pyrethrum extract in 19 parts of kerosene (v/v)	Flit pump or hand operated fogging (micro discharge) machine	Used for Indoor space spray
2	Malathion	Technical Malathion	5 parts of Tech. Malathion in 95 parts of Diesel (v/v)	Thermal fogging machine	Outdoor thermal fogging. Watch for wind conditions before fogging

Precautions during Indoor Fogging

Following precaution to be taken during indoor fogging:

1. The fog should reach all possible resting sites of vectors. This is particularly (very crucial for Dengue vectors).

2. All electric switches, heating and cooking equipment must be put off and allowed to cool before undertaking fogging.

3. All water containers and foodstuffs etc. must be well protected from exposure to fog.

4. During the fogging all animals and human beings should remain outside and stay there for at least 30 minutes after spray. After that house should be ventilated by opening windows and doors.

5. All doors and windows should be closed before spray and kept closed for at least 30 minutes for maximum efficacy.

6. Begin fogging from most interior place and move away from fog while covering the area.

7. Fog must be dry before release into the household. It can be checked by placing machine at ground and observing the area immediately in front of nozzle. It should not be wet.

Precautions during Outdoor Fogging

The climatic conditions, traffic, movement of operator etc affect fogging efficiency. Following precautions are to be taken in outdoor fogging:

1. The doors and windows of all shelters should be kept open to allow penetration of fog so that it could reach all possible resting sites of vectors.

2. During the raining and high wind velocity fogging should not be done.

3. The distance between shelters and operator should be kept minimum to allow fog penetration in shelters.

4. The down wind side of the spray area should be treated first, working systematically from downwind to upwind.

5. For avoiding exposure to spray cloud the dead end road must be sprayed only on the way out.

Environmental Management

The management strategies and their impact may be short-term or long-term, and may require community involvement and multi sectoral action. In Malaysia, breeding of the malaria vector *An. maculatus* in streams has been effectively controlled by periodic flushing by means of small dams with siphons and sluice gates, where in Indonesia, *An. sundiacus* has been controlled by changing the salinity of the breeding habitats as an environmental management approach to vector control. Such types of project adopted in several countries, including India. Experience with methods of environmental management applied in the per-DDT era should now be reviewed and their utility explored because they may have been effective in many situations, but were discontinued because of the expectations that malaria could be eradicated with DDT.

Development Activities

Environmental change created during development activities can increase the risk of malaria. Agricultural practices, irrigation development or road and building construction projects, etc increases the vector problem and are the indirect cause of health problems.

Housing : domestic and peridomestic environment.

The quality of housing (the design, structure and construction material) and its location in relation to breeding sites influence mosquito entry, resting habits and human vector contact. Incomplete houses with open walls, wide or unscreened eaves, open windows, doors, and no ceilings favour the entry of mosquitos. Mud or unplastered wall with cracks and crevices and thatched roofs or walls also provide favoured resting sites for mosquitoes. Unscreened water-storage containers and long-standing water bodies located in and near houses provide mosquitoes with breeding habitats.

Forest Ecosystems

Mining and agriculture at the forest fringe and deforestation favour human contact with efficient vectors. Therefore, forest ecosystem is a high-risk environment for malaria transmission.

Use of Insecticide-Treated Mosquito Nets

Mosquitoes transmit malaria, filaria, dengue, chikungunya, yellow fever and certain other diseases. Therefore, insecticide treated nets be used for protecting humans from mosquitoes and diseases they transmit. Pregnant women, babies and young children are at the greatest risk of dying of malaria and other diseases. Ordinary untreated mosquito nets provide limited physical barrier to mosquitoes to bite. However, they may still bite through the net. Thus, for better and effective protection, keeping away mosquitoes and killing them mosquito nets treated

234 | *Mosquito Diversity and Control*

with insecticides are must. An insecticide-treated mosquito net also kills or keeps away house hold insects such as bed bug, cockroaches, sandflies, houseflies, fleas, crickets, etc.

Preventive Control

Legislative Measures

Suitable laws and byelaws should be prepared and implemented for regulating storage/utilization of water by communities, various agencies and avoidance of mosquitogenic conditions at construction sites, factories.

1. *Model civic byelaws*: Under this act fine/ punishment is imparted. These measures should be strictly enforced by Municipal Corporations in India.

2. *Building Construction Regulation Act*: Building byelaws should be made for not allowing stagnation of water vis-à-vis breeding of mosquitoes. This includes under ground tanks, mosquito proof buildings, designs of sunshades, porticos, etc. The owners/builders should deposit a fee for controlling mosquitogenic conditions at site.

3. *Environmental Health Act (HIA)*: Suitable byelaws should be made for the proper disposal/storage of junk, discarded tins, old tyres and other debris, which can withhold rain water.

4. *Health Impact Assessments*: Appropriate legislation should be formulated for mandatory HIA prior to any development projects/major constructions.

Health Education for Community Mobilization and Inter-Sectoral Covergence

Special campaigns should be arranged through mass media including local vernacular newspapers/magazines, radio and TV, especially using local cable networks as well as outdoor publicity like hoardings, miking, drum beating, rallies, etc. Health education materials should be developed and widely circulated in the form of posters, pamphlets, handbills, hoardings. Inter-personal communication through group meetings, traditional/folk media should be made.

Biological Control of Mosquito Larvae Using Fishes

1. Introduction

As early as 1903 fishes have been used in public health for control of mosquitoes. The top water minnow or mosquito fish *Gambusia affinis* and *Poecilia reticulata* have received the most attention as a mosquito control agents in India and abroad.

Advantages of Use of Fish

☆ These fishes are self-perpetuating after its establishment and continuous to reduce mosquito larvae for long time.

☆ The cost of introducing larvivorous fishes is relatively lower than that of chemical control.

☆ Use of fishes is ecofriendly/environment friendly method of control.

☆ Larvivorous fishes (*Gambusia* and *Poecilia*) prefer shallow water where mosquito larvae can breed constantly.

Larvivorous Fish should have following features:

1. Small in size.
2. Surface feeders and carnivorous.
3. Able to survive in the absence of mosquito larvae.
4. Easy to rear.
5. Able to withstand a wide range of temperature and light intensity.
6. Hardy and able to withstand transport and handling.
7. Insignificant/useless as food for other predators.
8. Preference for mosquito larvae over other types of food available.

Gambusia

Biocontrol potential of *Gambusia*:

1. A single full grown fish eats about 100 to 300 mosquito larvae per day.
2. *Gambusia* is a surface feeder, hence it is suitable for feeding on both anophelines and culicines.
3. It frequents the margins of the water container, pond or other ground water collections, except where there is dense vegetation at the margins of the water body.
4. It is small and inedible fish and tolerate salinity.
5. It is sturdy in transportation and does not require any specialized equipment.
6. It survives in new places (water bodies) and multiplies easily. After release when it becomes well

established in a water body, the fish can survive in good numbers for years together without special care.

Poecilia reticulata (Guppy)

P. reticulata an exotic fish was introduced in India in 1910. It is easy to rare or reproduce quickly and prolifically. It is very widely utilized for biological control of mosquitoes. The male measures for 3 cm long, while the female measures about 6 cm in body length. The Guppy lives for 4 + 1 years and acts as very good biocontrol agent of mosquitoes which feed on mosquito larvae.

It can survive in all types of water bodies. This hardy fish tolerates high degree of pollution with organic matter. It has developmental temperature range from 24°C to 34°C for breeding and water pH range from 6.5 to 9.0 for survival. However, it can not survive in cold water below 10°C

The guppy matures within 90 days. Each ovary contains 100 to 160 eggs. The female gives birth to young ones in broods of 5 to 7 at a time for every four weeks, about 50 to 200 young ones are released by the female. It breeds throughout the year at about four weeks interval after maturity. In warmer climate it may breed from April to November.

Biocontrol potential of *Poecilia* is due to following characters:

1. A single fish eats about 80 to 100 mosquito larvae in 24 hours. However, it comparatively less efficient than *Gambusia affinis*.

2. It is a surface feeder and negotiates margins of ponds more easily.

3. It is highly carnivorous and parents or older brood may eat up their own young ones. Hence, weeds are required in the water for sheltering/hiding young ones.
4. Tolerant capacity against handling and transportation.
5. No specialized equipment for transportation.
6. Survives and reproduces into new water bodies and found in the habitat even after many years.

Genetic Control

Chemosterilants

Mosquitoes can be effectively controlled by using some chemosterilants especially alcalating agence like tepa 0.1 per cent and metapa 0.06 per cent and alpholate 0.6 per cent.

Irradiating Males

Another method of sterilization is the radio sterilization of males with gamma irradiation causing a distortion of chromosomes and there by making them sterile. Large number of such sterile males are introduced into high density mosquito population areas thereby producing unfertile eggs.

Cytoplasmic Incompatability

Normal offspring was prevented in crosses between alien strains (of different geographical origin) owing to the incompatible factors in the egg cytoplasm. The sperms from the males enter the egg cell and even induce embryo formation but the sperm nucleus is prevented from uniting with the egg nucleous by plasmagenes in the cytoplasm.

References

Atwal, A.S. 1933. Agricultural pests of India and South east Asia. *Kalyani Publ.*, 1- 289 pp.

Baisas, F.F. and F.H. Dowell 1965. Key to the adult females and larval anopheline mosquitoes of the Philippines. *PACAI Epidem. Lab. Tech. Rept.* 1365. 52 pp., Illus.

Baker, F.C. 1936. A new species of Orthopodomia alba sp.n. (Diptera, Culicidae). *Proc. ent. Soc. Wash.* 38 : 1-7, 1 pl.

Banks, C.S. 1906a. A list of Philippine Culicidae, with description of some new species. *Philipp. J. Sci.* 1 : 977-1005.

Banks, C.S. 1906b. Four new Culicidae from the Philippines. *Philipp. J. Sci.* 4 : 545-551.

Banks, C.S. 1914. A new Philippine Malaria mosquito. *Philipp. J. Sci.* (D), 9 : 405-407.

Barr, A.R. 1957. A new species of Culiseta (Diptera : Culicidae) from North America. *Pro. ent. Soc. Wash.* 59 : 163-167. Illus.

Barraud, P.J. 1923a. A revision of the Culicidae mosquitoes of India. Part-V. Further notes on the genera Stegomyia, Theo. and Finlaya Theo. with descriptions of new species. *Indian J. Med. Res.*, 11 : 224-228. 3pls.

Barraud, P.J. 1923b. Two new species of Culex (Diptera, Culicidae) from Assam. *Indian J. Med. Res.* 11 : 507-509.

Barraud, P.J. 1923c. A revision of the culicine mosquitoes of India. *Indian J. Med. Res.* Vol. xi : 971-976.

Barraud, P.J. 1924a. A new mosquitoes from Kashmir. *Indian J. Med. Res.* 11 : 967-968, Illus.

Barraud, P.J. 1924b. Four new mosquitoes from the Western Himalayas. *Indian J. Med. Res.* 11 : 999-1006. Illus.

Barraud, P.J. 1924c. A revision of the culicine mosquitoes of India-XV. The Indian species of the subgenus Lophoceratomyia (Theo, Edw.) including one new species. *Indian J. Med. Res.* 12 : 39-45, 2 pls.

Barraud, P.J. 1924d. A revision of the culicine mosquitoes of India. Part XII. Further descriptions of Indian species of Culex L. including two new species. *Indian J. Med. Res.* 11 : 1259-1274, 3 pls.

Barraud, P.J. 1924e. A revision of the Culicine mosquitoes of India Part XIII. Further descriptions of Indian species of Culex L. including three new species. *Indian J. Med. Res.* 11 : 1275-1282, 2 pls.

Barraud, P.J. 1924f. A revision of the Culicine mosquitoes of India XIV. The Indian species of the subgenus

Culiciomyia (Theo.) Edw., including one new species. *Indian J. Med. Res.* 12 : 15-22, 1 pls.

Barraud, P.J. 1924g. A revision of the Culicine mosquitoes of India Part XV. The Indian species of the subgenus Lophoceratomyia (Theo.) Edw., including two new species. *Indian J. Med. Res.* 12 : 39-46, 2 pls.

Barraud, P.J. 1924h. A new mosquitoes from Kashmir and the North-West Frontier province. *Indian J. Med. Res.* 12 : 73-84.

Barraud, P.J. 1926. A revision of the Culicine mosquitoes of India. Part XVIII. The Indian species of Uranotaenia and Harpagomyia, with descriptions of five new species. *Indian J. Med. Res.* 14 : 523-532, 1 pl.

Barraud, P.J. 1927a. A revision of the Culicine mosquitoes of India XIX. The Indian species of Aedomyia and Orthopodomyia with description two new species. *Ibid.* XIV, pp, 523-532.

Barraud, P.J. 1927b. A revision of the Culicine mosquitoes of India Part XX. The Indian species of Armigers (including Leicesteria) with descriptions of two new species. *Indian J. Med. Res.* 14 : 523-548, 2 pls.

Barraud, P.J. 1927c. A revision of the culicine mosquitoes of India Part XXI. Description of new species of Aedimorphus and Finlaya and notes on Stegomyia albolineata (Theo.). *Indian J. Med. Res.* 14 : 549-544, Illus.

Barraud, P.J. 1927d. A revision of the Culicine mosquitoes of India Part XXII. The Indian species of the genus Taeniorhynchus (including Mansonioides with a description of one new species. *Indian J. Med. Res.* 14 : 549-554, Illus.

Barraud, P.J. 1928. A revision of the Culicine mosquitoes of India. The Indian species of the subgenera Skusea and Aedes, with description of eight new species and remarks on a new method for identifying the females of the subgenus Aedes. *Indian J. Med. Res.* 16 : 357-365, 8 pls.

Barraud, P.J. 1931a. Notes on some Indian mosquitoes of the subgenus Segomyia, with descriptions of new species. Indian J. Med. Res. 19 : 221-227, 1 pl.

Barraud, P.J. 1931b. Descriptions of eight new species of Indian culicine mosquitoes. *Indian J. Med. Res.* 19 : 609-416, 1 pl.

Barraud, P.J. 1931c. Notes on some Indian mosquitoes of the subgenus Stegomyia, with description of new species. *Indian J. Med. Res.* 19 : 221-227, 1 pls.

Barraud, P.J. 1931d. Notes on some Indian mosquitoes of the subgenus Stegomyia with description of two new species. Indian J. Med. Res. 20 : 101-109.

Barraud, P.J. 1932. A new species of Anopheles pinjaaurensis n.sp. *Rec. Malar. Surv. India* 3 : 353-355, Illus.

Barraud, P.J. 1934. The fauna of British India, including Ceylon and Burma. *Taylor and Franas : London* 1-436.

Barraud, P.J. and S.R. Christophers 1931. On a collection of anopheline and described new Culicine mosquitoes from Siam. *Rec. Malar. Surv. India.* 2 : 269-285.

Baruah, I.; N.G. Das and J. Kalita 2007. Seasonal prevalence of malaria vector in Sonitpur district of Assam. *J. Vect. Borne. Dis.* 44 : 149-153 pp.

Basu, P.C. 1958. A note on malaria and filariasis in Andaman and Nicobar Island. *Bull. Nat. Soc. Ind. Med. Mosq. Dis.*, 6 : 193 – 206.

Bates, M. 1940. The Nomenclature and taxonomic status of the mosquitoes of the *Anopheles malulipensis* complex. *Ann. ent. Soc. Am.* 33 : 343-356.

Belkin, J.N. 1945. *Anopheles nataliae*, a new species from Guodal canal. *J. parasit.* 31 : 315-318. Illus.

Belkin J.N. 1962. The Mosquitoes of the South Pacific (Diptera, Culicidae) [sic]. Vols. I and II. *University of California Press, Berkeley and Los Angeles.*

Belkin, J.N. 1968. Mosquito studies (Diptera, Culicidae) VII. The Culicidae of New Zealand. *Contributions of the American Entomological Institute (Ann Arbor)*, 3(1), 1– 182.

Belkin, J.N. 1970. Mosquito studies (Diptera : Culicidae). XXI. The Culicidae of Jamaica. *Contr. Am. ent. Inst* 6(I) : 1-458.

Belkin, J.N. and R.J. Schlosser 1944.A new species of Anopheles from Solomon *Island. J. Wash. Acad. Sci.* 34 : 268-273, Illus.

Belkin, J.N., R.X. Schick and S.J. Heinemann 1968. Mosquito studies (Diptera : Culicidae)-XI Mosquito originally described from Argentina, Bolivia, Chile, Paraguay, Peru and Uruguay. *Contr. Am. ent. Inst.* 4(1) : 9-29.

Belkin, J.N., Heinemann, S.J. and Page, W.A. 1970. The Culicidae of Jamaica (Mosquito Studies. XXI). *Contributions of the American Entomological Institute (Ann Arbor)*, 6(1): 1–458.

Belkin, J.N.; R.X. Schick and S.J. Heinemann 1971. Mosquito studies (Diptera : Culicidae) xxx. Mosquitoes originally described from Brazil. *Contr. Am. ent. Inst.* 7(5) : 1-64.

Bentley, C.A. 1902. A new Anopheles mosquito in Tezpur, Assam. *Indian Med. Gaz.* 37 : 15-16, Illus.

Bhatia, M.L., B.L. Wattal, M.L. Mannen and N.L. Kerla 1958. Seasonal prevalence of anophelines near Delhi.

Bigot, J.M. 1861. Trois dipteres nouveaux de la corse. *Ann. Soc. ent. Fr.* (4) 1 : 227-229.

Bohart, R.M. 1946.A key to the Chinese Culicine mosquitoes *U.S. Navmed.* 961 : 1-23.

Bohart, R.M. 1950. A new species of Orthopodomyia from California (Diptera : Culicidae). *Ann. ent. Soc. Am.* 48 : 399-404, Illus.

Bohart, R.M. 1956. New species of mosquitoes from the Southern Ryukyu Islands. *Bull. Brooklyn ent. Soci.* 51 : 29-34, Illus.

Bram, R.A. 1967a. Contributions to the mosquito fauna of Southeast Asia. II. The genus Culex in Thailand (Diptera: Culicidae). *Contributions of the American Entomological Institute (Ann Arbor),* 2(1), iii + 1–296.

Bram, R.A. 1967b. Classification of Culex subgenus Culex in the New World (Diptera: Culicidae). *Proceedings of the United States National Museum,* 120(3557): 1–122.

Bram, R.A. 1968. A redescription of Culex (Acalleomyia) obscurus (Leicester). *Proceedings of the Entomological Society of Washington,* 70: 52–57.

Brunetti, E. 1912.Annotated catalogue of Oriental Culicidae supplement *Rec. Indian Mus.* 4 : 403-517.

Brunetti, E. 1914. Critical review of genera in Culicidae. *Rec. Indian Mus.* 10 : 15-73.

Busvine, J.R. 1980. Insect and hygience. The biology and control of insect pests of Medical and domestic importance 521-527 pp.

Cagampang-Romos and R.F. Darsie 1970. Illustrated key to the *Anopheles* mosquitoes of the Philippine islands. *USAF Fiftn. Epidemilogical Hight, PACAF Tech. Report,* 70 :1- 49 pp., Illus.

Carpenter, S.J. 1966. Review of recent literature on mosquitoes of North America. Calif. *Vector Views,* 15 : 71-88.

Carpenter, S.J. and W.J. Lacasse 1955. Mosquitoes of North America (North of Mexico) *Uni. California Press,* vi + 360 pp. illus., 127 pls.

Carter, H.F. 1910. A new Anopheline from South Africa. *Entomologist* 43 : 237-238, Illus.

Carter, H.E. 1911. A new mosquito from Uganda. *Bull. ent. Res.* 2 : 37-38, Illus.

Chamnarn Apiwathnasorn, 1986. A list of mosquito species in Southeast Asia. Museum and Reference Centre.

Chen, C.Y. 1972. Studies on morphology of the Cinarium in Culicine mosquitoes-I. Eight species of Culicines common in the Taipei area, *Taiwan, Formosan med. Ass. J.* 71 : 282-291, Illus.

Chow, C.Y. 1949. The identification and distribution of Chinese Anopheline mosquitoes. *J. nat. malar. Soc.* 8 : 121-131.

Christophers, S.R. 1911. Notes on mosquitoes II A new Anopheline paludism simla No.2 : 64-68, 1 pl.

Christophers, S.R. 1915. The male genitalia of Anopheles. *Indian Journal of Medical Research*, 3: 371–394 + 6 pls.

Christophers, S.R. 1924. Provisional list and reference catalogue of the Anophelini. *Indian Medical Research Memoirs*, 3 : 1–105.

Christophers, S.R. 1933. The fauna of British India, including Ceylon and Burma. *Taylor and Franas : London* 1-360.

Christophers, S.R. 1960. *Aedes aegypti* (L.) the yellow Fever Mosquito. *Cambridge Univ. Press, Cambridge, Engl*; 739 pp.

Christophers, S.R. and P.J. Barraud 1931. The eggs of Indian Anopheles with descriptions of the hitherto undescribed eggs of a number of species. *Rec. Malar. Surv. India* 2 : 161-192, 5 pls.

Click, J.I. 1992. Illustrated key to the female Anopheles of South-western Asia and Egypt (Diptera : Culicidae). *Mosq. Syst.* 24(2) : 125-153.

Colless, D.H. 1948. The Anopheline mosquitoes of northwest Borneo. *Proc. Linn. Soc. N.S.W.* 73 : 71-119, Illus.

Colless, D.H. 1960. Some species of Culex (Lophoceraomyia) from New Guinea and adjacent island, with description of four new species and notes on the male of Culex Fraudatrix Theobald (Diptera, Culicidae). *Proc. Linn. Soc. N.S.W.* 84 : 382-390. Illus.

Colless, D.H. 1965. The genus Culex subgenus Lophoceraomylai in Malaya (Diptera : Culicidae) *J. Med. Ent.* 2 : 261-307, Illus.

Coquillett, D.H. 1901. Three new species of Culicidae. *Can. Ent.* 33: 258-260.

Conquillett, D.H. 1902. Six new species of Culex from America. *Can. Ent.* 34 : 292-299.

Coquillett, D.H. 1906. Five new Culicidae from the West Indies. *Can. Ent.* 38 : 60-62.

Cooper R. D., Waterson D. G. E., Frances S. P. Beebe N. W. and Sweeney A. W. 2002. Speciation and distribution of the member of the *Anopheles punctulatus* (Diptera – Culicidae) group in Papua New Guinea. *J. Med. Ento.* 39 : 16–27.

Danilov, V.N. 1989. Two new Afrotropical subgenera of the mosquito genus Culex (Diptera: Culicidae). *Entomologischeskoe Obozrenie*, 68: 790–797 (in Russian).

Darsie, R.F. and D.A. Shayer 2004. *Culex* (Culex) *declarator* a mosquito species new to Filorida. *J. Am. mosq. control Assoc. Vol.* 20(13) : 224-227.

Das, S.C., N.G. Das and I. Baruah 1984. Mosquito survey in Meghalaya. *Indian J. Pub. Health* xxviii (3) : 147-150 pp.

De Meillon, B. 1947. The Anophelini of the Ethiopian geographical region. *Publ. S. Afr. Inst. Med. Res.* 10 : 1-272.

Delfinado, M.D. 1966. The Culicine mosquitoes of the Philippines, tribe Culicini (Diptera, Culicidae). *Memoirs of the American Entomological Institute*, 7: 1–252.

Dobrotworsky, N.V. 1965. The Mosquitoes of Victoria. *Melbourne*, 237 pp.

Draft. 2007. National Biodiversity Action Plan, Government of India, Ministry of Environment and forests, pp 1-97.

Dyar, H.G. 1905. Remark on genitalic genera in the Culicidae. *Proc. ent. Soc. Wash.* 7 : 42-49 Illus.

Dyar, H.G. 1917. A second note on the species of Culex of the Bahamas (Diptera : Culicidae). *Insec. mencit.* 5 : 183-187.

Dyar, H.G. 1920. A collection of mosquitoes from the Philippine Islands (Diptera : Culicidae). *Insec. Inscit. Menst.* 8 : 175-186.

Dyar, H.G. 1928. The mosquitoes of the Americas. *Carnegie Institute of Washington Publication,* 387, 1–616, 123 pls. + errata and addend.

Edwards, F.W. 1912. A key to the Australian species of Ochlerotatus (Culicidae). *Ann. Mag. nat. Hist.* (8) 9: 521-527.

Edwards, F.W. 1913. New synonymy in Oriental Culicidae. *Bull. ent. Res.* 4 : 221-242.

Edwards, F.W. 1916. Eight new mosquitoes in the British Museum collection. *Bull. ent. Rec.,* 6 : 357-364 Illus.

Edwards, F.W. 1920. Catalogue of oriental and South Asiatic Nematocera. *Rec. Indian Mus.* 17 : 1-300.

Edwards, F.W. 1921a. A revision of the mosquitoes of the palaearctic Region. *Bull. ent. Res.* 12 : 263-251, Illus.

Edwards, F.W. 1921b. A synonymic list of the mosquitoes hitheto recorded from Sweden, with key for determining the genera and species. *Ent. Tidskr.* 42 : 46-52.

Edwards, F.W. 1923. A revision of the Culicine mosquitoes of India Part-III. Notes on certain Indian species of the genus Finlaya and descriptions of new species. *Indian J. Med. Res.* 11 : 214-219.

Edwards, F.W. 1923. Oligocene mosquitoes in the British Museum; with a summary of our present knowledge concerning fossil Culicidae. *Quarterly Journal of the Geological Society of London*, 79: 139–155, 1 pl.

Edwards F.W. 1932. Genera Insectorum. Diptera, Fam. Culicidae. *Fascicle*, 194. Desmet-Verteneuil, Brussels.

Edwards, F.W. 1941. Mosquitoes of the Ethiopian Region III. Culicidae adults and pupae, *British Museum (Nat. Ural History), Longon*, 499 pp., Illus., 4 pls.

Evans, A. M.1931. A new subspecies of *Anopheles funestus* Giles, from Southern Rhodesia. *Ann. trop. Med. parasit*, 25 : 545-549.

Evans, A.M. 1938. Mosquitoes of the Ethiopian Region II Anophelini adults and early stages. *British Museum (Natural History), London* 404 pp. Illus.

Foote, R.H. 1954. The larvae and pupae of the mosquitoes belonging to the Culex subgenera. Melanoconion and Mochlostyrax. *Tech. Bull. U.S. Dep. Agric.* 1091: 1- 126 pp.

Foote, Richard H. and Cook, David R., 1959. Mosquitoes of Medical Importance. U.S. Dept. Agric., *Agric. Handbook.* 152: 1-158 pp.

Gelfand, H.M. 1954. The Anopheline mosquitoes of Liberia. *W. Afr. Med. J. (N.S.)* 3 : 80-88, Illus.

Giles, G.M. 1899. A description of the Culicidae employed by Major R. Ross, I.M.S. in this investigation on Malaria. *J. trop. Med.* 2 : 62-65, Illus.

Giles, G.M. 1901a. Six new species of Culicidae from India. *Entomologist,* 34 : 192-197.

Giles, G.M. 1901b. Description of four new species of Anopheles from India. *Ent. mon. Mag.* 37 : 196-198.

Giles, G.M. 1904. Notes on some collections of mosquitoes and C., received from the Philippine Islands and Angola with some incidential remarks upon classification. *J. trop. Med.* 7 : 365-369, Illus.

Gillies, M.T. and B. Meillon 1968. The Anophelinae of Africa South of the Sahara. *Publs. S. Afr. Inst. med. Res.* 54 : 343 pp.

Girhe, B. E. 2001a. Biodiversity of mosquitoes from Kolhapur district, Maharashtra *M. Phil. Thesis, Shivaji University,* pp.1-150.

Girhe, B. E. and T. V. Sathe 2001b. Incidence of Malaria in Kolhapur district, Maharashtra, *Nat. Sym. Devt. Environ. and Human cands,* No.3.6: 36 pp.

Girhe, B.E. and T.V. Sathe 2001c. On a new species of the genus Aedes Meign (Diptera : Culicidae) from *India. J. Ady. Zool.* 22 (1) : 46-47.

Godfray H. C. J. 2002. Challenges for taxonomy. *Nature.* 417 : 17-19.

Graham, H. 1905. Mosquitoes intermediary vector of Diseases. *Serv. Publ. Sch. Publ. Health Sydney* 3, 20 pp., Illus.

Grabham, M. 1906. Four new Culicidae from Jamaica, West Indies. *Can. Ent.* 38 : 311-320, Illus.

Gubler D. J. 1996. World wide distribution of dengue in 1996. *Dengue Bulletin*, 20:1-12.

Gutsevich, A.V., Monchadskii, A.C. and Shtakel Berg. A.A. 1970. Fauna of U.S.S.R. Vol. III No.4, Mosquitoes. Family : Culicidae Leningrad : Zoological Institute 384 pp. (In Russian).

Hackett, L.W. and D.J. Lewis 1935. A new variety of *Anopheles maculipennis* in Southern Europe. *Riv. Malariol.* 14 : 377-383.

Horsfall, W.R. 1955. Mosquitoes: Their bionomics and relation to disease. 723 pp. New York.

Hussainy, Z.H. 1981. Distribution records of culicine mosquitoes of Bastar district Madhya Pradesh, India (Diptera : Culicidae*) J. Bombay Nat. Hist. Soc.*, 77 : 277-284.

Idem, 1910. Monograph of the Culicidae of the World, 5 pp. 19: 50–51.

Jaal Z, Macdonald WW. 1993. The ecology of Anopheline mosquitos [*sic*] in northwest coastal Malaysia: Larval habitats and adult seasonal abundance. *Southeast Asian J Trop Med Public Health*, 24:522-529.

James, S.P. and W.G. Liston 1904. A monograph of the Anopheles mosquitoes of India Vi + 132 pp., XV uncolored + XV coloured pls. Culcutta.

James, S.P. and W.G. Liston 1905. Some new mosquitoes from Ceylon. *J. Bombay nat. Hist. Soc.* 16 : 237-250, 2 pls.

James, S.P. and W.G. Liston 1910. A new classification of the Anopheline. *Rec. Indian Mus.* 4 : 95-109, 4 pls.

James, S.P. and W.G. Liston 1911. A monograph of the Anopheles mosquitoes of India. vi + 132 pp. XV uncolered + XV coloured pls. Calcutta.

Jepson, W.F.; A. Moutia and C. Courtois 1947. The malaria problem in Mauritius : the bionomics of Mauritian anophelines. *Bull. ent. Res.* 38 : 177-208, Illus.

Joshi, V.; R.C. Sharma; M. Singhi; H. Singh; K. Sharma; Y. Sharma and S. Adha 2005. Entomological studies on malaria in irrigated and non-irrigated areas of Thar desert, Rajasthan, India. *J. Vect. Borne. Dis.* 42: 25-49 pp.

Kanojia, P.C.; P.S. Shetty and G. Geevarghese 2003. A long term study on vector abundance and seasonal prevalence in relation to the occurrence of Japanese encephalitis in Gorkhapur district, Uttar Pradesh. *Indian J. Med. Res.* 117 : 104-110.

Keshishian, M. 1938. *Anopheles sogdianus* sp.nov. A new species of the Anopheles mosquito A. sogdianus sp. nov. in Tadjikistan. *Med. Parazit.* (Moscow) 7: 888-896.

King, W.V. and H. Hoogstraal 1946. Two species of Aedes (Finlaya) from New Guinea (Diptera : Culicidae) *Pro. ent. Soc. Wash.* 42 : 37-38.

Knight, K.L. 1947. The *Aedes* (Finlaya) *chrysolineatus* group of mosquitoes (Diptera : Culicidae). *Ann. ent. Soc. Am.* 40 : 624-649, illus.

Knight, K.L. 1969. A new species of the genus Aedes, subgenus Finlaya Theobald, from India (Diptera : Culicidae). *J. Kansas. ent. Soc.* 42 : 382-386, Illus.

Knight, K.L. 1978. Supplement to a catalog of the mosquitoes of the world (Diptera: Culicidae). Thomas

Say Foundation, *Entomological Society of America*, 6(Suppl.), 1–107.

Knight, K.L. 1947. The *Aedes* (Finlaya) *chrysolineatus* group of mosquitoes (Diptera : Culicidae). *Ann. ent. Soc. Am.* 40 : 624-649, Illus.

Knight, K.L. 1969. A new species of the genus Aedes, subgenus Finlaya Theobald, from India (Diptera : Culicidae). *J. Kansas. ent. Soc.* 42 : 382-386, Illus.

Knight, K.L. 1978. Supplement to a catalogue of the mosquitoes of the world (Diptera: Culicidae). Thomas Say Foundation, Entomological Society of America, 6(Suppl.), 1–107.

Knight, K.L. and J.L. Laffoon 1946. The Oriental species of the Aedes (Finlaya) Kochi group (Diptera : Culicidae). *Trans. Am. ent. Soc.* 72 : 203-225, Illus.

Knight, K.L. and W.B. Hull. 1951. The Aedes mosquitoes of the Phillippine Islands-I. Keys to species. Subgenera Mucidus, Ochlerotatus and Finlaya (Diptera : Culicidae). *Pacif. Sci.* 5 : 211-251, Illus.

Knight, K.L. and W.B. Hull 1952. The Aedes mosquitoes of the Philippine Island II. subgenera Skusea, Christophersicmyia, Geoskusea, Rhinoskusea and Stegomyia (Diptera : Culicidae). *Pacif. Sci.* 6 : 157-189, Illus.

Knight, K.L. and Stone, A. 1977. A catalogue of the mosquitoes of the world (Diptera: Culicidae). *Second edition. Thomas Say Foundation*, 6: 1–611.

Knowles, R. and White R. 1927. Malaria, its investigation and control. Thacker Spink and Co., Calcutta.

Knowlton, N., and L. A. Weigt. 1998. New dates and new rates for divergence across the Isthmus of Panama. *Proc. R. Soc. Lond. B Biol. Sci.* 265: 2257-2263.

Korke, V.T. 1928. Observations on Filariasis in some area in British India I and II. *Indian Jour. Med. Res.*, 16: 717-732.

Korke, V.T. 1932. Observations on Filariasis in some areas in British India, VIII. *Indian Jour. Med. Res.*, 20 : 335-339.

Lacasse, W.J. and S. Yamaguti, 1950. Mosquito fauna of Japan and Korea, 268 pp. App. I. The female terminalia of the Japanese mosquitoes, 7 pp., App. II. Organization and function of Malaria survey detachments, 213 pp. 34th Edition, illus. off. Surgeon, 8th U.S. Army, Kyoto, Honshu.

Laird, M. 1956. Studies of mosquitoes and fresh water ecology in the South Pacific. *Bull. R. Soc. N. Z.* 6 : 213 pp.

Lane, J. 1953. Neotropical Culicidae 112 pp., Illus. Sao Paulo, Brazil.

Lee, K.W. and J.C. Lein 1970. Pictorial key to the mosquitoes of Korea. *WHO/VBC/70* 196. 7 PP., Illus.

Linnaeus, C. 1758. Systema naturae per regna tria naturae. Edition 10, Vol.1, 824 pp. Holmiae.

Linnaeus, C. 1762. Zweyter Theil, enthalt Beschreibuhgen verschiedener wichtiger Naturalien. pp. 267-606.

Ludlow, C.S. 1905. Mosquito notes No.3. *Can. Ent.* 37 : 94-102: 129-135.

Lynch Arribalzaga 1891. Dipterologia argentina. *Revta Mus. La plata* 1 : 345-377.

M.R.C. 2006. Bionomic of Malaria Vector in India. *J. Vector Biology*, Vol. 3: 19-41 pp.

Mahendra Jagtap, and T. V. Sathe 2008. Role of Intensified mass surveillance campaign in malaria problematic area of Sangli district. *Perspectives in Animal Ecology and Production* (ISBN10 81-7035-563-X) vol.5: 14-24.

Mallet J and Willmott K. 2003. Taxonomy: Nenaissance or Tower of Babel? *Trends in Ecology and Evolution*. 18 : 57 – 59.

Mani, M. A. 1982. General Entomology. *Oxf. and IBH pub.* C0. Pvt. Ltd. 1-900 pp.

Marsh, F. 1933. A new species of Anopheles (Myzomyia group) from South-West persia. *Stylops* 2 : 193-197.

Mattingly, P.F. 1952. The subgenus Stegomyia (Diptera : Culicidae) in the Ethiopian Region, I.A. Preliminary study of the distribution of species occurring in the West African subregion with notes on taxonomy and bionomics. *Bull. Brit. Mus. (nat. Hist.)* (B) 2 : 233-304, Illus.

Mattingly, P.F. 1953. The sub-genus Stegomyia (Diptera : Culicidae) in the Ethiopian Region II. Distribution of species confined to the East and South Africa subregion. *Bull. Brit. Mus.* (nst. Hist.) (B) 3 : 1-65, Illus.

Mattingly, P.F. and Knight, K.L. 1956. The mosquitoes of Arabia I. Bulletin of the British Museum (Natural History) *Entomology*, 4: 91–141.

Mattingly, P.F. 1958. The Culicine mosquitoes of the Indomalayan Area. Part-III. Genus Aedes Meigen, subgenera paraedes Edwards, Rhinoskusea Edwards and Cancrdedes Edwards. 61 pp., illus. *Brist. Mus. nat. Hist., London.*

Mattingly, P.F. 1959. The Culicine mosquitoes of the Indomalayan Area. Part-IV. Genus Aedes Meigen, subgenera skusea Theobald, Diceromyid Theobald, Geoskusea Edwards and Christophersiomyia Barraud 61 pp., Illus. *Brist. Mus. nat. His., London.*

Mattingly, P.F. 1961. The Culicine mosquitoes of the Indomalayan Area. Part-V. Genus Aedes Meigen, subgenera Mucidus Theobald, Ochlerotatus Lynch Arribalzaga and Neomeldniconion Newstead. 62 pp., illus. *Brit. Mus. nat. Hist., London.*

Mattingly, P.F. 1965. The Culicinae Mosquitoes of the Indomalayan Area VI. Genus Aedes Meigen Subgenus Stegomyia Theobald (Groups A, B and D).

Mattingly, P.F. 1969. The Biology of Mosquito-borne Disease. George Allen and Unwin Ltd, London.

Mattingly, P.F.1970. Contributions to the mosquito fauna of Southeast Asia. – VI. The genus Heizmannia Ludlow in Southeast Asia. *Contributions of the American Entomological Institute* (Ann Arbor), 5(7): 1–104.

Mattingly, P.F. 1971. Contributions to the mosquito fauna of Southeast Asia. – XII. Illustrated keys to the genera of mosquitoes (Diptera, Culicidae). *Contributions of the American Entomological Institute* (Ann Arbor), 7(4): 1–84.

Mattingly, P.F. 1981. Medical entomology studies – XIV. The subgenera Rachionotomyia, Tricholeptomyia and Tripteroides (Mabinii Group) of genus Tripteroides in the Oriental Region (Diptera: Culicidae). *Contributions of the American Entomological Institute* (Ann Arbor), 17(5): ii + 1–147.

Mattingly, P.F. and K.L. Knight 1956. The mosquitoes of Arabia-I. *Bull. Brit. Mus.* (nat. Hist.) (B) 4 : 89-496, illus.

McAlpine, J.E. 1979. Diptera H.V. Danks (ed.) Canada of its insecta fauna. Mem. *Ent. Soc. Can.* 108-573.

Meigen, J.W. 1818. Systematische Beschreibung der bekannten europaischen zweiflugeligen Insekten. Vol. 1 xxxvi +333 pp. 1-11 nymphes d' anophelines (4 e Memoire) *Arch. Inst. Pasteur Alger* 12 : 29-76, Illus.

Michener, C.D. 1944. Differentiation of females of certain species of Culex by the Cibarial armature. *J.N.Y. ent. Soc.* 52 : 263-266, Illus.

Mishra, B.G., 1956. Malaria in Northeast frontier Agency (India) *Indian J. Malariol.* 10 : 331-347.

Mithalyi, F. 1963. Biting Mosquitoes of Hungary 229 pp. Budapest : Akademiai Kiado (In Hungarian).

Mulligan, H.W. and I. M. Puri 1936. Description of *Anopheles* (Anopheles) *habibi* n. sp. from Quetta, Baluchistan. *Res. Malar. Surv. India.* 6 : 67-71.

Murthy, S.U.; K.S. Sai; D.V. Satyakumar; K. Siriram; K. Madhusudhan Rao; D. Krishna and B.S. Murthy 2002a. Relative abundance of *Culex quinquefasciatus* (Diptera : Culicidae) with reference to infection and infectivity rate from rural and urban areas of East and West Godavari districts. *J. Trop. Med. Pub. Health.* Vol. 33(4) : 110-119.

Murty, S.U.; V.R. Satyakumar; K. Siriram; K. Madhusudhan Rao; T. Gopalsingh; N. Arunachalam and

P. Philisamuel 2002b. Seasonal prevalence of *Culex vishnui* subgroup, the major vector of Japanese encephalitis in a endemic district of Andhra Pradesh, India. *J. Am. mosq. control Assoc. USA* 18(14) : 290-293pp.

Nagpal, B.N. and V.P. Sharma 1983a. Mosquitoes of Andaman Island. *Indian J. Malariol.* 20 : 7-13 pp.

Nagpal, B.N.; Y. Kumar; U. Sharma and V.P. Sharma 1983b. Mosquitoes of Nainital Terai (U.P.). *Indian J. Malariol.* 20 : 129-135 pp.

Nagpal, B.N. and V.P. Sharma 1987. Survey of mosquito fauna of North eastern region of India. *Indian J. Malariol.* 24: 143-149 pp.

Nagpal, B.N. and Sharma V.P. 1995. Indian Anophelines. Science Publishers, Inc. Lebanon, New Hampshire, USA.

Nair, K.S.S. and Mathew, G. 1993. Diversity of insects in Indian forests – the state of our knowledge.

Neveu-Lemaire, M. 1902. Classification de la famille des Culicidae. *Mem. Soc. Zool. Fr.* 15 : 195-227.

Nguygen, D.M., Trans D.N., Ralphe, H., Jonna E. and Y.M. Liston 2000. A new species of the Hyrcanus group of Anopheles, subgenus Anopheles, A secondary vector of malaria in coastal area of Southern Vietam. *J. Am. mosq. control Assoc.* 16 (3): 189-198.

Pal, T.R. and R.K. Dutta 1992. Anophelines (Diptera : Culicidae) of three districts (East Kameng, Lower Subansiri and Upper subansiri) of Arunachal Pradesh and their perspective impact on Human and nonhuman Host. *Res. Zool. Surv. India,* 91 (9) : 189-202.

Pemola, N. and R.K. Jauhari 2006. Climatic variable and Malaria incidence in Dehradun, Uttaranchal, India. *J. Vect. Borne Dis.* 43: 21-28 pp.

Peyton, E.L. and S. Ramalingam 1988. *Anopheles* (Cellia) *nemophilous* a new species of the Leucosphyrus Group from peninsular Malaysia and Thailand (Diptera : Culicidae) *Mosq. Syst.* 20 : 272-299.

Poinar, Jr., J.O., Zavortink, T.J., Pike, T. and Johnston. 2000. *Paleoculicis minutus* (Diptera: Culicidae) n. gen., n. sp., from Cretaceous Canadian amber, with a summary of described fossil mosquitoes. *Acta Geologica Hispanica*, 35: 119–128.

Postiglione, M., C.B. Smiraglia; A. Lavagnino; C. Gokberk and C. Ramsdale 1970. A preliminary note on the occurrence in Turkey of the Subalpinus from the A. maculipennis Complex. *Riv. Parassit.* 31 : 155-158.

Pringle, G. 1954. The identification of the adult Anopheline mosquitoes of Iraq and neighbouring territories. *Bull. Endem. Dis.* 1 : 53-76.

Puri I.M. 1929. A new tree-hole breeding Anopheles from South India. *Anopheles sintoni* and a revised description of the larva of Anopheles culiciformis Cogil. *Indian J. Med. Res.* 17 : 397- 404, 1 pls.

Puri, I.M. 1935. Schematic table for the identification of the Indian Anophelines mosquitoes. Part-I, Adult; Part-II Larvae. *Rec. Mal. Surv. Ind.* 5 : 265-273.

Puri, I.M. 1941. Synoptic table for the identification of the anopheles mosquitoes of India. Health. Bull. No. 10. Malaria Bureau No.2 Govt. of India, Calcutta.

Puri, I.M. 1947. The practical application of D.D.T. for Malaria control in rural and urban areas in India.

Puri, I.M. 1948. The distribution of Anopheline mosquitoes in India, Pakistan, Ceylon and Burma. Additional records. 1936-1947 *Ind.J.Malariol.*, 2: 67 – 107.

Puri, I.M. 1960. Synoptic table for the identification of full grown larvae of the Indian Anopheline mosquitoes. *Health Bull. No. 16*, Govt. of India, Calcutta.

Qutubuddin M. 1960. Mosquito studies *in the Indian subregion Part I Taxonomy – A brief review. Pacific* Insects, 2(2):133-147.

Rahman, M., M. D. Elias and H. Mahmud ul Ammen. 1977. Bionomics of *An. balabacensis* in Bangladesh and its relation to malaria. *Bangladesh J. Zool.*, 5:1-237.

Rajavel, A.R. 1996. Taxonomic notes on Culex (Lophoceraomyia) infantulus Edwards (Diptera : Culicidae). Based on its new distribution records in Pondicherry, South India. *Entomon.* 21 (1) : 43-48.

Rajavel, A.R.; R. Natarajan; K. Vaidyanathan and A. Munirathinam 2000. Seasonal incidence of *Aedes* (Rhinoskusea) *portonovoensis* in a mangrove forest of South India. *J. Am. mosq. control Asso.* Vol. 16(4) : 1-8pp.

Rajavel, A.R.; A. Munirathinam R. Natarajan and K. Vaidyanathan 2001. Species diversity of mosquitoes (Diptera : Culicidae) in Mangrove ecosystem in South India. *Entomon.* 26(3 and 4) : 271-277.

Rajavel, A.R.; R. Natarajan and K. Vaidyanathan 2004. A check-list of mosquitoes (Diptera : Culicidae) of

Pandicherry, India with notes on new area records. *J. Am. mosq. control. Asso.* Vol. 20(3) : 228-332.

Rajavel, A. R., R. Natarajan, K. Vaidyanathan, and V. P. Soniya. 2005. A list of the mosquitoes housed in the mosquito museum at the Vector Control Research Centre, Pondicherry, India. *J. Am. Mosq. Control Assoc.* 21: 243-251.

Rajavel, A.R.; R. Natarajan and K. Vaidyanathan 2005a. Mosquito of the mangrove forests of India : Part-I Bhitarkanika, Orissa. *J. am. mosq. control. Asso.* vol. 21(2) : 131-135.

Rajavel, A.R.; R. Natarajan and K. Vaidyanathan 2005b. Mosquito collections in the Jeypore hill tracts of Orissa, India, with notes on three new country records, *Culex* (Lophoceraomyia) *pilifemoralis*, *Culex* (Lophocerao-myia) *wilferedi* and *Heizmannia* (Heizmannia) *chengi*. *J. Am. mosq. control. Asso.* Vol. 21(2) : 121-127 pp.

Rajavel, A.R.; R. Natarajan; K. Vaidyanathan and V.P. Soniya 2007. Mosquitoes of the Mangrove forests of India : Part-5 Ghorao, Goa and Vikhroli, Maharashtra. *J. Am. mosq. control Assoc.* Vol. 23(3) : 343-345 pp.

Rajput, K.B. and S. M. Kulkarni 1990. Records of Culicine mosquitoes from Bastar district (Madhya Pradesh) Indian (Diptera : Culicidae), Part-I. Genus Toxorhynchites, Tripteroides, Uranotaenia and Orthopodomyia. *Rec. Zool. Surv. India,* 87(1) : 83-88.

Rajput, K.B. and S.M. Kulkarni 1991. Records of Culicine mosquitoes from Basatar district (Madhya Pradesh), India (Diptera : Culicidae), Part-II. Genus Aediomyia,

Armigers, Coquillettidia, Heizmannia, Mansonia and Mimomyia. *Rec. Zool. Surv. India*, 89 : (1-4) : 251-255.

Rajput K.B. and T.K. Singh 1992a. Records of mosquitoes (Diptera : Culicidae), from Manipur. Genus Culex. *Rec. Zool. Surv. India*, 91 (3-4): 108-118.

Rajput K.B. and T.K. Singh 1992b. Records of mosquitoes (Diptera : Culicidae) from Manipur Genus Aedes. *Rec. Zool. Surv. India*, 91(3-4) : 293-302.

Rajput K.B. and T.K. Singh, 1992c. Records of mosquitoes (Diptera : Culicidae) from Manipur Genus Armigeres and Heizmannia. *Rec. Zool. Surv. India*, 91(3-4) : 241-286.

Ralph E. Harbach. 2007. The Culicidae (Diptera): a review of taxonomy, classification and phylogeny. *Zootaxa*, 1668: 591-638.

Rao R.T. 1975. Arboviruses and their vectors in India. *Indian J. Med. Res.*, 63 : 1219-1234.

Rao, R. 1984. The Anophelines of India. Second edition. Malaria Research Centre, Indian Council of Medical Research, New Delhi 518 pp.

Reid, J.A. 1953. The *Anopheles hyrcanus* group in Southeast Asia (Diptera : Culicidae) Bull. *Entomol. Res. 44 :* 5-76.

Reid, J.A. and K.L. Knight 1961a. Anopheline mosquitoes as vector of animal in Malaya. *Ann. trop. Med. parasit.,* 55 : 180-186.

Reid, J.A. and Knight, K.L. 1961b. Classification within the subgenus Anopheles (Diptera, Culicidae). *Annals of Tropical Medicine and Parasitology*, 55: 474–488.

Reid, J.A. 1966. Anopheline mosquitoes of Malaya and Borneo studies from the Institute for Medical Research, Kaulalampur No. 32: 22 : 529 pp.

Reid, J.A. 1968. Anopheline mosquitoes of Malaya and Borneo. Studies from the Institute for Medical Research Malaya, 31: 1–520.

Reinert, J.F. 2000. New classification for the composite genus Aedes (Diptera: Culicidae: Aedini), elevation of subgenus Ochlerotatus to generic rank, reclassification of the other subgenera, and notes on certain subgenera and species. *J. of the Am. Mos.Con. Ass.*, 16: 175–188.

Reuben, R. 1967. Aedes (Diceromyia) ramachandrai n. sp. (Diptera : Culicidae) from South India. *J. Med. Entomol.* 4(2) : 234-236.

Reuben, R.; V. Thenmozhi; P.P. Samuel; A. Gajanan and T.R. Mani 1992. Mosquito blood feeding patterns as a factor in the epidemiology of Japanese Encephalitis in Southern India. *Am. J. Med. Hyg.* 46(6) : 654-663 pp.

Reuben, R., S. C. Tewari, J. Hiriyan, and J. Akiyama. 1994. Illustrated keys to species of Culex (Culex) associated with Japanese encephalitis in Southeast Asia (Diptera: Culicidae). *J. Am. Mosq. Control Assoc.* 26: 75-96.

Ross, H.H. 1951. Conflict with Culex. *Mosquito News*, 11: 128–132.

Roy, D.N. and Brown, A.W.A. 1970. Entomology 3rd Edn. *Bangalore Press Bangalore*, 855 pp.

Rush, D.V. 1780. Dengue Fever in Philadelphia. Bull. *Brookiyan ent. Soc.* 40 : 20-30, Illus.

Salem, H.H. 1938. The mosquito fauna of sinai peninsula (Egypt) with a description of two new species. *Publ. Fao. Med. Egypt.* Univ. 16 : 1-31.

Sallum, M.A. Peyton, E.L. and R.C. Wilkerson 2005. Six new species of the *Anopheles leucosphyrus* group reintertion of *Anopheles eleyans* and vector implications. *J. Medical and Veternary Entomology*, Vol. 19(2): 158-199.

Sathe, T.V. and Girhe, B.E. 2001. Biodiversity of Mosquitoes (Order : Diptera) in Kolhapur district, Maharashtra. *Riv. Di. Parassitologia*, 18 (LXVII-3): 189-194.

Sathe, T. V. and Girhe, B. E. 2002. Mosquitoes and Diseases, *Daya Publ. House, New Delhi*. pp 1-96.

Sathe T. V. 2006. Biological control of mosquitoes. *Proc. rec. trends in Malaria studies, Pune* pp.11-17.

Sathe T. V. and B. P. Tingare 2006. Biodiversity of mosquitoes in Solapur city, Maharashtra. *Nat. proc. workshop Biological control Insect pest.* pp.8.

Sathe, T.V. and B.P. Tingare 2007. On a new species of the genus Anopheles Meigen (Diptera : Culicidae) from India. *Indian J. Environ and Ecoplan*, 14 (1-2) : 61-64.

Sathe, T. V. and Mahendra Jagtap. 2008. Three decades trend of malaria from Sangli district of Maharashtra, India. *Perspectives in Animal Ecology and Production* (ISBN10 81-7035-563-X) 5: 25-37.

Sathe, T. V. and Mahendra Jagtap. 2009. Tree hole breeding and resting of mosquitoes in Western Ghats, Maharashtra. *J.Exp.Zool. India.* 12(2): 365-367.

Sathe, T. V. and Tingare, B. P. 2010a. Mosquito Biodiversity, *Mang. Publ. Delhi.* pp 1-227.

Sathe, T. V. and Mahendra Jagtap. 2010b. Biodiversity of Anopheline mosquitoes from Western Ghats of Maharashtra (India). *Biospectra.* 5(2): 215-218.

Sathe, T. V. Asavari Sathe and Mahendra Jagtap. 2010c. Mosquito Borne diseases, *Mang. Publ. Delhi.* pp 1-342.

Satishkumar, B.Y. and V.A. Vijayan 2005. Mosquito fauna and breeding habitats in the rural areas of Mysore and Mandy districts, Karnatak State, India. *Entomon.* 30 (2) : 123-129.

Sazanova, O.N. 1958. The key for the identification of females of the genus Aedes (Diptera : Culicidae) of the forest zones of the USSR. *Riv. Ent. URSS,* 37 : 741-752, Illus.

Scanlon J.E.; E.L. Peyton and D.J. Gould 1968. An annotated checklist of the Anopheles of Thailand. Thai. *Natl. Sci. paper Fauna Ser.* No.2, 35 pp., Illus.

Sen, R.N., R. Rajagopal and Chakraborty 1960. Observations on the seasonal prevalence of adult Anophelines near Dhanbad. *Indian J. Malariol.* 14 : 23-54.

Sen, S.K. Johm, V.M.; Krishnan, K.S. and Rajgopal, R. 1973. Studies on malarial transmission in Trip district, Arunachal Pradesh (NEFA). *J. Com. Dist.* 5 : 98-110.

Senevet, G. 1947. Le genre Culex en Afrique de Nord.1 Clef de determination des larvaes. *Arch. Inst. Pasteur Alger.* 25 : 212-213.

Senevet, G. and L. Andarelli 1954a. Le genre Culex Afrique du Nord III-Les adultes. *Arch. Inst. Pasteur Alger.* 32 : 36-70, Illus.

Senevet, G. and L. Andarelli, 1954b Le genre Aedes en Afrique du Nord-I. Les larves. *Arch. Inst. Pasteur Alger.* 32 : 310-351, Illus.

Senior White, R. 1923. Catalogue of Indian insects, Part 2 - Culicidae, Calcutta, p. 26 pp.

Senior White, R. 1937. On malaria transmission in the Jeypore Hills Part-I. A year's dissection results. *Rec. mal. Surv. India* 7 : 47-75.

Senior White, R. 1943. On malaria transmission Ranchi Hazaribagh ranges including Ranchi Plateau. *J. Mal. Inst. Ind.* 5 : 207-231.

Service, M.W. 1962. Key to the West African Anophelini. *Acta. Tropica* 19 : 120-158.

Shahgudian, E.R. 1960. A key to the Anopheline of Iran. *Acta. Med Iran.* 3 : 38-48.

Shidrawi, G.R. and M.J. Gillies 1987. *Anopheles paltrinierii* n. sp. (Diptera : Culicidae) from the sultanate of Omen. *Mosq. Syst.* 19 : 201-211.

Shukla, R.P., S.N. Sharma and R.C. Dhiman. 2007. Seasonal prevalence of malaria vectors and its relationship with malaria transmission in three physiographic zones in Uttaranchal state, *India. J. Vect. Borne Dis.* 44 : 75-77.

Sicart, M. 1954. Contribution a 1' etude des nymphs du genre Culex en Tunisie. *Bull. Soc. Sci. nat. Tunis* 7 : 27-29.

Sirivanakaran, S. 1972. Contribution to the mosquito fauna of Southeast Asia-XIII. The genus Culex, subgenus Eumelanomyia Theobald in Southeast Asia and adjacent areas. *Contr. Am. ent. Inst.* 8(6) : 1-86, Illus.

Smart, J. 1965. A handbook for the Identification of Insects of medical importance 303 pp. 4th Edn. London : B.E. Mus (Nat. Hist.).

Smith, M.E. 1958. The Aedes mosquitoes of new England. Part-I. Key to adult females. *Bull. Brooklyn ent. Soc.* 53 : 39-47, Illus.

Smith, M.E. 1969. The Aedes mosquitoes of New England (Diptera : Culicidae) II. *Can. Ent.* 101 : 41-51.

Steward, C.C. and McWade, J.W. 1961. The mosquitoes of Ontario (Diptera : Culicidae) with keys to the species and notes on distribution. *Proc. ent. Soc. Ont.* 91 : 121-188.

Stone, A. 1945. A mosquito synonym (Diptera : Culicidae) *Proc. ent. Soc. Wash.* 47: 38-39.

Stone, A. 1961. A synoptic catalog of the mosquitoes of the world, suppl-I (Diptera : Culicidae). *Proc. ent. soc. Wash.* 63 : 29-52.

Stone, A. 1963. A synoptic catalog of the mosquitoes of the world, suppl-II (Diptera : Culicidae). *Proc. ent. soc. Wash.* 65 : 117-140.

Stone, A. 1967. A synoptic catalog of the mosquito of the world. Suppl. III (Diptera : Culicidae) *Proc. ent. soc. Wash.* 69 : 197-224.

Stone, A. 1970. A synoptic catalogue of the mosquitoes of the World Suppl. IV. (Diptera : Culicidae). *Proc. ent. Soc. Wash.* 72 : 137-171.

Stone, A. and G.H. Penn 1948. A new subgenus and two new species of the genus Culex L. (Diptera : Culicidae). *Proc. ent. Soc. Wash.* 50 : 109-120, Illus.

Stone, A. and Barreto, P. 1969. A new genus and species of mosquito from Colombia, Galindomyia leei (Diptera, Culicidae, Culicini). *Journal of Medical Entomology*, 6: 143–146.

Stone, A. and M.D. Delfinado 1973. A catalogue of the mosquitoes of the world (Diptera : Culicidae). *The Thomas say Foundation Entomological Society of America* 611 pp.

Stone, A., K.L. Knight and Helle Starcke, 1969. A synoptic catalogue of the mosquitoes of the world (Diptera : Culicidae). The Thomas say foundation, *Ent. Soc. Am.* Vol. 6. 358 pp.

Stone, A., Knight, K.L. and Starcke, H. 1959. A synoptic catalog of the mosquitoes of the world (Diptera, Culicidae). *Thomas Say Foundation*, 6: 1–358.

Strickland, C. 1916. A new species of protanopheline from Malaya, *Myzorhynchus hunteri* sp.nov. *Indian J. Med. Res.* 4 : 263-270 illus.

Strickland, C. and K.L. Chowdhury 1927. A new species of Anopheline. *Anopheles psedojamesi* common in Bengal. *Indian Med. Gazette* 62 : 240-243, Illus.

Tanaka, K. M. 1979. A revision of the adult and larval mosquitoes of Japan and Korea (Diptera : Culicidae). *Contri. Amer. Entomol. Inst.* 16 : 1-17.

Tanaka, K. 2003. Studies on the pupal mosquitoes of Japan (9). Genus Lutzia, with establishment of two new subgenera, Metalutzia and Insulalutzia (Diptera, Culicidae). *Japanese Journal of Systematic Entomology*, 9: 159–169.

Tautz D., Arctander P., Minelli A., Thomas R. H. and Vogler A. P. 2003. A plea for DNA taxonomy. *Trends Ecol. Evol.* 18 : 70 – 74.

Taylor, R.M. 1967. Catalogue of Arthropod - borne viruses of the world. *Public. Health Service Publication* No. 1760. 898 pp. Washington D.C., U.S. Govt. Printing Press.

Tewari, S.C., J. Hiriyan and R. Reuben 1987. Survey of the Anopheline fauna of the Western Ghats in Tamil Nadu, India. *Indian J. Malariol.* Vol. 24: 21-28.

Tewari, S.C. and J. Hiriyan 1990. Description of the male larva and pupa of *Aedes* (Diceromyia) *kanarensis* (Diptera : Culicidae) *Mosq. Syst.* Vol. 22 : 41-46.

Tewari, S.C. and J. Hiriyan 1991. Description of the new species of Aedes (Rhinoskusea) from South India. *Mosq. Syst.* Vol. 23(2) : 123-131.

Tewari, S.C. and J. Hiriyan 1992. Description of two new species of Aedes (Diceromyia) from South India (Diptera : Culicidae). *Mosq. Syst.* Vol. 24 : 154-175.

Tewari, S.C. and J. Hiriyan 1994. Description of the female pupa and larva of Aedes (Paraedes) Barraud and the pupa and larva of Aedes (Paraedes) Menoni (Diptera : Culicidae) *Mosq. Syst.* 26(3) : 105-115.

Theobald, F.V. 1901a. A Monograph of the Culicidae or mosquitoes. Vol. 1: 1- 424 pp., Illus. London.

Theobald, F.V. 1901b. A Monograph of the Culicidae or mosquitoes. Vol. 2, viii + 391 pp., Illus. London.

Theobald, F.V. 1902. The classification of Anopheles. *J. trop. Med.* 5 : 181-183, Illus.

Theobald, F.V. 1903a. A monograph of the Culicidae or mosquitoes. Vol. 3: 359 pp., illus., 17 pls. London.

Theobald, F.V. 1903b. Report on a collection of mosquitoes or culicidae etc. from Gambia and descriptions of new species. *Mem. Lpool. Sch. trop. med.* 10 (App): i-xi,2 pls.

Theobald, F.V. 1904a. A new Culicidae genus from Uganda. *J. trop. Med.* 7 : 17-18, Illus.

Theobald, F.V. 1904b. New Culicidae from the Federated Malay States. *Entomologist* 37 : 12-15, 36-39, 77-78.

Theobald, F.V. 1905a. Genera Insectorum Diptera, Fam. Culicidae. *Fascicle* 26: 50 pp., 2 pls. Belgium.

Theobald, F.V. 1905b. A new genus of Culicidae. *Entomologist*, 38 : 52-56, Illus.

Theobald, F.V. 1905c. A catalogue of the Culicidae in the Hungarian National Museum, with descriptions of new genera and species, *Ann. hist. nat. Mus. hung.* 3 : 61-120, illus., 4 pls.

Theobald, F.V. 1907. A monograph of the Culicidae or mosquitoes. Vol. 4, 639 pp., illus., 16 pls.

Theobald, F.V. 1910. A Monograph of the Culicidae or Mosquitoes. Vol. V. British Museum (Natural History), London.

Theobald, F.V. 1910. Second report on the collection of Culicidae in the Indian Museum, Calcutta, with descriptions of new genera and species. *Rec. Indian Mus.* 4 : 1-33, 3 pls.

Theobald, F.V. 1911. A new genus and two new species of Culicidae from the Sudan. *Report of Welcome Research Laboratories, Gordon College, Khartoum* 4B : 151-156.

Thurman, D.C. 1954. Ayurakitia, a new genus of mosquito from northern Thailand (Diptera : Culicidae). *J. Wash. Acad. Sci.* 44 : 197-200.

Tilak, R.K.; K. Duttagutta and A.K. Verma 2006. Vector Databank in the Indian Armey Forces. *MJAFI*, Vol. 64, No. 1: 36-39 pp.

Van Bortel. W., Ttrung H. D., Manh N. D., Roelants P., Verle P. and Coosemans M. 1999. Identification of two species within the *Anopheles minimus* complex in Northern Vietnam and their behavioural divergences. *Trop. Med. And Int. health.* 4 : 257 – 265.

Van Someren, E.C.C., Teesdale, C. and M. Furlong, 1955. The mosquitoes of the Kenya coast; records of occurrence, behaviour and habitat. *Bull. ent. Res.* 46 : 463-493.

Vanden Assem J. and J. Bonne 1964. New Culicidae, a synopsis of vectors, pests and common species. *Zool. Bijdri.* 6: 136 pp.

Vander Wulp, F.M. 1884. Note on exotic Diptera. Part-I. *Notes Leyden Mus.* 6 : 248-256.

Ward, R.A. 1984. Second supplement to "A Catalog of the Mosquitoes of the World" (Diptera: Culicidae). *Mosquito Systematics*, 16: 227–270.

Ward, R.A. 1992. A catalogue of the mosquitoes of the world (Diptera : Culicidae) *Mosq. Syst.* 24 : 177-230.

Wattal, B.L. and N.L. Kalra 1961. Regionwise pictorial key to the female Indian Anopheles. *Bull. Nat. Sac. Ind. Mal. Mosq. Dis.*, 9 : 85-138.

Weyer, F. 1954. Bestimmungsschluessel fuer die Anopheles Weibchen and larven in Europa, Nordafrika und Westasian 2nd suppl. ed. *Merkbl. Nocht. Inst. Schiffsu Tropenkrankh* 12 : 3-38, Illus.

WHO. 2002. The World Health Report. Reducing risks, Promoting Healthy Life.

Wilkerson, R.C. and D. Strickman 1990. Illustrated key to the female Anopheline mosquitoes of Central America and Mexico. *J. Am. Mosq. Control. Assoc.* 6 : 7-34.

Zaim, M. and Z. Javaherian 1991. Occurrence of *Anopheles culicifacies* species A in Iran. *J. Am. Mosq. Control Assoc.* 7 : 324-326.

Zavortink, T.J. 1968. Mosquito studies (Diptera, Culicidae) VIII. A prodrome of the genus Orthopodomyia. *Contributions of the American Entomological Institute* (Ann Arbor), 3(2): 1–220.

Zavortink, T.J. 1970a. The treehole Anopheles of the New world. Contr. *Am. ent. Inst.* 5(2) : 1-35.

Zavortink, T.J. 1970b. Mosquito studies (Diptera, Culicidae) XIX. The treehole Anopheles of the world. *Contributions of the American Entomological Institute* (Ann Arbor), 5(2): 1–35.

Zavortink, T.J. 1972. Mosquito studies (Diptera, Culicidae) XXVIII. The New World species formerly placed in Aedes (Finlaya). *Contributions of the American Entomological Institute* (Ann Arbor), 8(3): 1–206.

Index